**Computation: Computability,
Similarity and Duality**

Hong Jia-wei
Beijing Computer Institute, China

Computation: Computability, Similarity and Duality

Pitman, London

John Wiley & Sons, Inc., New York, Toronto

PITMAN PUBLISHING LIMITED
128 Long Acre, London WC2E 9AN

A Longman Group Company

© Hong Jia-wei 1986

First published 1986

Available in the Western Hemisphere from
John Wiley & Sons, Inc.
605 Third Avenue, New York, NY 10158

ISSN 0268-7534

British Library Cataloguing in Publication Data

Hong, Jia-wei
   Computation: computability, similarity
   and duality.
   1. Computer arithmetic
   I. Title
   519.4′028′5   QA76.9.C62

ISBN 0-273-08720-7

Library of Congress Cataloging in Publication Data

ISBN 0-470-20387-0

All rights reserved. No part of this publication may be reproduced,
stored in a retrieval system, or transmitted, in any form or by any
means, electronic, mechanical, photocopying, recording and/or
otherwise, without the prior written permission of the publishers.
This book may not be lent, resold, hired out or otherwise disposed
of by way of trade in any form of binding or cover other than that
in which it is published, without the prior consent of the publishers.

Reproduced and printed by photolithography
in Great Britain by Biddles Ltd, Guildford

Contents

PREFACE

PRELIMINARIES 1
I.   Notation 1
II. Alphabet and Language 2
III. Graph 3

PART ONE : FINITE AUTOMATA AND COMPUTABILITY 5
1. Finite Automata 5
   §1 Deterministic Finite Automata 5
   §2 Non-deterministic Finite Automata 6
   §3 Regular Language and Regular Expression 9
   §4 Properties of Regular Languages 10
   §5 Two-way Deterministic Finite Automata 14

2. The Turing Machine 17
   §6 Computability 17
   §7 The Turing Machine 18
   §8 Multitape Turing Machines 20
   §9 Turing Machine Combination 22
   §10 The Universal Turing Machine 23

3. Recursive Functions 27
   §11 Definitions and Examples 27
   §12 The Arithmetization of Turing Machines 32
   §13 Computing Recursive Functions by Turing Machines 36
   §14 Recursive and Recursively Enumerable Languages 40

PART TWO : DETERMINISTIC SIMILARITY		44
4.	Complexity of the Turing Machine	44
	§15 Computational Complexity	44
	§16 Resources of the Multitape Turing Machine	46
	§17 Basic Relationships and Properties	59
	§18 Log-Space Transform Machine	65
	§19 Log-Space Constructible Graphs and Nice Pair of Functions	68
	§20 The Concept of Similarity	76
5.	Multi-Index Random Access Machine	80
	§21 Definition	80
	§22 Examples	88
	§23 RAM Simulating TM	94
	§24 LSTM Simulating RAM	96
6.	Vector Machines	102
	§25 Definition	102
	§26 Examples	106
	§27 Matrix Transpose and Word Projection	117
	§28 VM Simulating LSTM	125
	§29 The Similarity Between VM, RAM and TM	133
7.	Other Parallel Computational Models	136
	§30 Uniform Circuits	136
	§31 Uniform Aggregates	142
	§32 PRAM	144
	§33 The Similarity Between PRAM, UA, UC and Other Models	147
	§34 Some Remarks on Similarity	151
PART THREE : COMPUTATIONAL TYPES AND DUALITY		159
8.	Logical Computational Types	159
	§35 Non-deterministic Turing Machines and Non-deterministic Vector Machines	159
	§36 The Logical Computational Types for TM	165
	§37 The Classification of Logical Computational Types	169
	§38 Alternating Turing Machines	175

9.	General Computational Types	181
	§39 Reference Machine and The First Similarity Theorem	181
	§40 General Computational Types	185
	§41 Restrictions	188
	§42 The Third Similarity Theorem and Complexity Classes	194
	§43 The Majority and Random Types	197
	§44 Complete Problems	205
10.	The Duality Between Parallel Time and Space	210
	§45 The Boundary Theorem and Transform Theorem	210
	§46 The Symmetry Theorem and Restriction Theorem	212
	§47 The Complexity of Formal Proof	217
REFERENCES		228
INDEX		232

Preface

Computation is one of the oldest as well as one of the newest topics. Among the vast amounts of literature about computation, this book will find its own way to unify computational models and develop Turing-Church's thesis further.

This book composed of three parts. Part 1 (Chapters 1, 2 and 3) is a brief self-contained review, which discusses finite automata and classical computability theory. The material selected is as basic as possible, and the proof as simple as possible. The reader who is familiar with these materials can go to Part 2 directly.

Part 2 consists of Chapters 4, 5, 6 and 7. In Chapter 4, multitape Turing machines (TM) and computational complexities are introduced to the reader. In Chapters 5 and 6, random access machines (RAM) and vector machines (VM) are discussed, and their similarity with TM is proved. These three models are developed in full so that senior students and junior graduate students can understand the similarity theory for deterministic computational models. Chapter 7 provides three more parallel computational models which are proved to be similar to TM. Another four models and further developments are given in the exercises and remarks.

Part 3 consists of Chapters 8, 9 and 10. Chapters 8 and 9 are intended to unify various computational types. Especially non-deterministic, alternating, majority and random types, and their relations are considered. In Chapter 10, the duality between parallel time and space is discussed. This part can be treated as a separate text for further reading by graduate students.

This book is self-contained, but readers are recommended to complete the exercises in order to understand the text better.

The outline for this book was reported as a lecture at the 21st FOCS Symposium in 1980, entitled "On Similarity and Duality of Computation", based mainly on work I had done when I was visiting Toronto. Many valuable ideas were suggested by my friends A. Borodin, S. Cook, P. Dymond and C. Rackoff. In 1982, I wrote a lecture of 11 chapters for a theoretical seminar in Beijing Computer Institute and, in 1983, wrote a clearer version with the great help of S.W. Tang, for a summer school in Beijing. In 1985, I wrote it once again for *Research Notes in Theoretical Computer Science*, with continual encouragement from R. Book and A. Rosenberg, and helped by S.W. Tang, H.A. Wang and X.F. Liu. I would like to thank them all faithfully.

Hong, Jia-wei

Beijing Computer Institute,
China

Preliminaries

I. NOTATION

In this book, sets are denoted by capital letters, while its elements are denoted by lower-case letters. The total number of elements in set A is denoted by $|A|$. $\emptyset$ is the null set.

The union, intersection and difference of two sets are represented by $\cup$, $\cap$ and $-$ respectively.

$$A \cup B = \{x | x \in A \text{ or } x \in B\},$$
$$A \cap B = \{x | x \in A \text{ and } x \in B\},$$
$$A - B = \{x | x \in A \text{ and } x \notin B\}.$$

Notation $A \subseteq B$ means that A is a subset of B, while $A \subset B$ means $A \subseteq B$ and $B - A \neq \emptyset$.

The Cartesian product of A and B are defined by

$$A \times B = \{(a,b) | a \in A, b \in B\}.$$

Define $A^1 = A$, $A^{n+1} = A^n \times A$. Use 2^A to represent the power set of A:

$$2^A = \{B | B \subseteq A\}.$$

Use A^B to represent the set of all mappings from B to A.

A relation R on set A is a subset of $A \times A$. For two elements, $a, b \in A$, that a and b have relation R means $(a,b) \in R$, denoted by aRb.

The positive closure R^+ of a relation R is a relation defined by: aR^+b if and only if there are $a = a_1, a_2, \ldots, a_n = b$ ($n \geq 2$) such that $a_i R a_{i+1}$ ($i = 1, 2, \ldots, n-1$). The closure R^* of relation R is defined by: aR^*b if $a = b$ or aR^+b.

For any real number x, $\lfloor x \rfloor$ is the unique integer satisfying $\lfloor x \rfloor \leq x < \lfloor x \rfloor + 1$. $\lceil x \rceil$ is the unique integer satisfying $\lceil x \rceil - 1 < x \leq \lceil x \rceil$. The meaning of $[x]$ is the same as $\lfloor x \rfloor$.

II. ALPHABET AND LANGUAGE

An alphabet is a finite set, whose elements are called the symbols. Suppose that I is an alphabet. A string of a finite number (including 0) of symbols from I is a word over I. The word of length 0 is called the null word, denoted by Λ. The null word Λ is not a symbol in an alphabet but a string having no symbol, a word (empty word) over any alphabet. The set of all words in alphabet I will be denoted by I^*. For example, if $I = \{a,b\}$, then $I^* = \{\Lambda, a, b, aa, ab, ba, bb, aaa, \ldots\}$.

Suppose that $x = a_1 a_2 \ldots a_n$ and $y = b_1 b_2 \ldots b_m$ are two words in alphabet I. Then the word $w = xy = a_1 a_2 \ldots a_n b_1 b_2 \ldots b_m$ is called the concatenation of x and y. Obviously, we have

(1) $(xy)z = x(yz)$ $\quad x,y,z \in I^*$

(2) $\Lambda x = x \Lambda = x$ $\quad x \in I^*$.

If $xy = z$, then x is a prefix of z, and y is a suffix of z. Further, if we have $y \neq \Lambda$ (or $x \neq \Lambda$), then x (or y) is a proper prefix (proper suffix).

If $xyz = w$, then y is a subword of w. If we have $y \neq \Lambda$ and $y \neq w$ then y is a proper subword of w.

Suppose that $x = a_1 a_2 \ldots a_n$ ($a_i \in I$, $i = 1, 2, \ldots, n$), then $a_n \ldots a_2 a_1$ is the reversed word of x, denoted by x^r.

Let x be a word, $n \geq 0$. Define

$$x^0 = \Lambda$$

$$x^{n+1} = x^n \cdot x.$$

Obviously we have

(1) $x^n \cdot x^m = x^{n+m}$ $\quad (x \in I^*, n, m \geq 0)$

(2) $(x^n)^m = x^{nm}$ $\quad (x \in I^*, n, m \geq 0)$.

A subset of I^* is called a language over I.

Suppose that L_1 and L_2 are two languages over I. Define language

$$L = L_1 \cdot L_2 = \{xy \mid x \in L_1, y \in L_2\}$$

to be the concatenation of L_1 and L_2. Obviously we have

(1) $(L_1 \cdot L_2) \cdot L_3 = L_1 \cdot (L_2 \cdot L_3)$.

(2) $\{\Lambda\} \cdot L = L \cdot \{\Lambda\} = L$.

(3) $\emptyset \cdot L = L \cdot \emptyset = \emptyset$

(4) $(L_1 \cup L_2) \cdot L = L_1 \cdot L \cup L_2 \cdot L$

$L \cdot (L_1 \cup L_2) = L \cdot L_1 \cup L \cdot L_2$.

Thus $\{\Lambda\}$ and $\emptyset$ are essentially different. Furthermore, they are different from ␣ , the blank symbol.

Suppose that L is a language over I, $n \geq 0$. Define

$$L^0 = \{\Lambda\}$$

$$L^{n+1} = L^n \cdot L$$

$$L^* = \bigcup_{n=0}^{\infty} L^n$$

$$L^+ = \bigcup_{n=1}^{\infty} L^n.$$

L^* is the Kleene closure of L. L^+ is the positive Kleene closure of L. Obviously we have

$$(L^*)^* = L^*$$

$$L^+ = L \cdot L^* = L^* \cdot L.$$

Therefore $\emptyset^* = \{\Lambda\}$ but $\emptyset^+ = \emptyset$.

In the following we denote $\{w\}^*$ as w^*, and $\{w\}^+$ as w^+.

III. GRAPH

Suppose that V,E are finite sets and for every $e \in E$ there corresponds a unique $(u,v) \in V \times V$. Then $G = (V,E)$ is a directed graph. The elements in V are called the vertices. The elements in E are called the edges. If $(u,v) \in V \times V$ corresponds to $e \in E$, then u is the start point of e, and v is the end point of e.

The number of edges that have v as start (end) point is the fan-out (fan-in) number. A word $P = e_1 \ldots e_n$ over E satisfying that the end point of

e_i is the start point of e_{i+1} ($i = 1,2,...,n-1$), is called a path in the graph. The path is of length n. A path of length 0 is called a null path. Thus for every vertex v there is a null path from v to v. A non-null path having the same start point and end point is a cycle.

Suppose that G is a directed graph, I is a set, f is a mapping from E to I. Then (G,f) is an edge-assignment directed graph. f is called the assignment mapping. For $e \in E$, $f(e)$ is the assignment of e. Sometimes, G can be referred to as an I assignment directed graph. The assignment of a path $P = e_1 e_2 ... e_n$ is defined to be $f(P) = f(e_1)f(e_2)...f(e_n)$, which is a word over I.

Reversing the direction of all edges in a directed graph G, a new directed graph is obtained, the reversed graph G^r. If G is an I assignment graph, every edge in G^r keeps the original assignment. Obviously, P is a path in G if P^r is a path in G^r.

For each vertex in a directed graph, the set

$$L(v) = \{u | (v,u) \in E\}$$

is called the adjacency list of G.

For $G = (V,E)$, $V = \{v_1, v_2, ..., v_n\}$, the following matrix

$$A = \begin{pmatrix} a_{11} & \cdots & a_{1n} \\ \cdots & \cdots & \cdots \\ a_{n1} & \cdots & a_{nn} \end{pmatrix}$$

where

$$a_{ij} = \begin{cases} 1, & (v_i, v_j) \in E \\ 0, & (v_i, v_j) \notin E \end{cases}$$

is called the adjacency matrix of G.

The edge set of a directed graph is in fact a relation of V. If the relationship is symmetric, i.e., $(u,v) \in E$ iff $(v,u) \in E$, then the graph is an undirected graph.

Part one
Finite automata and computability
1 Finite automata

§1. DETERMINISTIC FINITE AUTOMATA

A deterministic finite automaton is a mechanism with a finite number of inner states. When it receives an input symbol it will respond with an output symbol and change to another state according to the current state and input symbol.

DEFINITION 1.1   A determinisitc finite automaton (DFA) is a 7-tuple:

$$M = (Q, I, U, \delta, \sigma, q_0, F)$$

where

    Q is a finite set of inner states;
    I is a finite set, the input alphabet;
    U is a finite set, the output alphabet;
    δ is a mapping from $Q \times I$ to Q, the state transition function;
    σ is a mapping from $Q \times I$ to U, the output function;
    $q_0 \in Q$, the initial state;
    $F \subseteq Q$, the set of final states.

Informally, a DFA has an FC, an input tape and an output tape. The FC, controlling a read head and a write head, is always in some state $q \in Q$ at any moment. The input tape and output tape are divided into many squares, each can store one symbol from I or U.

Initially, the FC is in the initial state q_0. The input, stored on the input tape, is a string $(a_1 a_2 \ldots a_n)$ of symbols in I where a_1 is the first input symbol; an empty word is on the output tape, i.e. all the squares on output tape are filled with blanks. The read head is pointing to the leftmost input symbol. According to the current inner state q_0 and the symbol a_1 scanned by the read head, the FC controls the write tape head to output a symbol $b_1 = \sigma(q_0, a_1)$ in the scanned square, enters a new state $q_1 = \delta(q_0, a_1)$ and the read/write heads automatically move one square to the right. Then, the DFA repeats the above process until all the input symbols have been treated. Finally, the FC enters a state q_n and produces an output word $b_1 b_2 \ldots b_n$ where

$$b_{i+1} = \sigma(q_i, a_{i+1})$$

$$q_{i+1} = \delta(q_i, a_{i+1}) \quad i = 0, 1, 2, \ldots, n-1.$$

For convenience, we extend the domain of σ and δ from $Q \times I$ to $Q \times I^*$:

$$\delta(q, \Lambda) = q,$$

$$\delta(q, wa) = \delta(\delta(q, w), a),$$

$$\sigma(q, \Lambda) = \Lambda,$$

$$\sigma(q, wa) = \sigma(q, w)\sigma(\delta(q, w), a), \text{ where } q \in Q, w \in I^*, a \in I.$$

For any given $w \in I^*$ if $\delta(q_0, w) \in F$, we say that the DFA accepts the input word w; otherwise, the DFA rejects the input word w.

For a simple DFA, we have an intuitive expression, the state-transition diagram. The diagram is a directed graph whose vertices correspond to the states of the DFA. For $q \in Q$, $a \in I$, if $\delta(q, a) = q'$, $\sigma(q, a) = b$, then there is an arc labelled a(b) from state q to state q' in the transition diagram. The initial state is generally indicated by an arrow.

When a DFA is used as a transducer, we often set $F = \phi$ (or $F = Q$). Thus, as a transducer a DFA can be described by a 6-tuple $(Q, I, U, \delta, \sigma, q_0)$.

The language accepted by M is defined by

$$L(M) = \{w \in I^* | \delta(q_0, w) \in F\}.$$

When a DFA is used as an accepter, we usually only write

$$M = (Q, I, \delta, q_0, F).$$

Then the DFA is a 5-tuple, and the arcs of its state transition diagram are labelled only by symbols from I.

§2. NON-DETERMINISTIC FINITE AUTOMATA

In the previous section, the language accepted by a DFA is the set

$\{w \in I^* |$ there is a path from q_0 to some $q \in F$ whose assignment is $w\}$. (2.1)

For each $q \in Q$ and each $a \in I$, $\delta(q,a)$ is unique.

Now we reduce the limits so that for each $q \in Q$ and each $a \in I$ there may be a finite number, perhaps zero, of arcs labelled a out from state q. We also specify one initial vertex q_0 and one accepting vertex set F, and define the language this graph accepts by (2.1). Thus, we get a new kind of language accepter.

DEFINITION 2.1 A non-deterministic finite automaton (NDFA) is a 5-tuple, $M = (Q, I, \delta, q_0, F)$, where Q, I, F and q_0 have the same meaning as for a DFA, but δ is a mapping from $Q \times I$ to 2^Q.

The function δ can be extended to a mapping from $Q \times I^*$ to 2^Q:

$$\delta(q, \Lambda) = \{q\},$$

$$\delta(q, wa) = \bigcup_{p \in \delta(q,w)} \delta(p, a) \quad (q \in Q, w \in I^*, a \in I),$$

and also can be extended to arguments in $2^Q \times I^*$ as follows:

$$\delta(S, a) = \bigcup_{q \in S} \delta(q, a) \quad (S \in 2^Q, a \in I) \quad (2.2)$$

$$\delta(S, \Lambda) = S,$$

$$\delta(S, wa) = \delta(\delta(S, w), a) \quad (S \in 2^Q, w \in I^*, a \in I).$$

The language accepted by M is:

$$L(M) = \{w \in I^* | \delta(q_0, w) \cap F \neq \emptyset\}$$

THEOREM 2.1 For each NDFA with k states, there exists an equivalent DFA with 2^k states. By 'equivalent', it is meant that they both accept the same language.

Proof: let $M = (Q, I, \delta, q_0, F)$ be an NDFA. Construct a DFA as follows:

$$M' = (2^Q, I, \delta', \{q_0\}, F'),$$

where

$$F' = \{S \in 2^Q | S \cap F \neq \emptyset\},$$

$$\delta'(S,a) = \bigcup_{q \in S} \delta(q,a).$$

According to formula (2.2), the function δ' is exactly the extension of the function δ, that is,

$$\delta'(S,a) = \delta(S,a) \quad (S \in 2^Q, a \in I).$$

But

$$w \in L(M') \leftrightarrow \delta'(\{q_0\},w) \in F' \leftrightarrow \delta(\{q_0\},w) \in F' \leftrightarrow \delta(q_0,w) \in F' \leftrightarrow \delta(q_0,w) \cap F \neq \emptyset \leftrightarrow w \in L(M).$$

Thus $L(M') = L(M)$.

Obviously, M' has 2^k states, where $k = |Q|$.

The model of the non-deterministic finite automaton can be extended to allow transitions on the empty input Λ.

DEFINITION 2.2 In Definition 2.1, if the function δ is a mapping from $Q \times (I \cup \{\Lambda\})$ to 2^Q instead of a mapping from $Q \times I$ to 2^Q, then we call M a non-deterministic finite automaton with Λ-moves, an NDFA with Λ-moves for short.

THEOREM 2.2 For each NDFA with Λ-moves, there is an equivalent NDFA without Λ-moves.

Proof: let $M = (Q,I,\delta,q_0,F)$ be an NDFA with Λ-moves. For each $q \in Q$, we define the Λ-CLOSURE of q as follows:

$$C_\Lambda(q) = \{p \in Q | \text{there is a path with assignment } \Lambda \text{ from } q \text{ to } p\}.$$

For each $S \subseteq Q$, define the Λ-CLOSURE of S as

$$C_\Lambda(S) = \bigcup_{q \in S} C_\Lambda(q).$$

Now define the map δ' from $Q \times I$ to 2^Q as follows:

For each $q \in Q$ and $a \in I$, $\delta'(q,a) = \delta(C_\Lambda(q),a)$.

Obviously,

$p \in \delta'(q,a) \leftrightarrow$ there is a path with a sequence of assignments Λ and finally an assignment a from q to p in M. (2.3)

Again, define

$$F' = \{q | C_\Lambda(q) \cap F \neq \phi\} \qquad (2.4)$$

$$M' = (Q,I,\delta',q_0,F').$$

Thus M' is an NDFA, which accepts L(M).

§3 REGULAR LANGUAGE AND REGULAR EXPRESSION

<u>DEFINITION 3.1</u> Let L be a language over the alphabet I. If L can be expressed with $\emptyset$, $\{\Lambda\}$, and $\{a\}(a \in I)$ through a finite number of operations of union $\cup$, concatenation, and Kleene closure *, then we call the language L a regular language (regular set).

<u>DEFINITION 3.2</u> Let I be an alphabet not containing the symbols (,), $\emptyset,\Lambda$, +, ·, and *.

(1) $\emptyset$ is a regular expression, denoting the language $\emptyset$. Λ is a regular expression, denoting the language $\{\Lambda\}$. For each $a \in I$, a is a regular expression, denoting the language $\{a\}$.

(2) If r and s are regular expressions denoting the languages R and S, respectively, then (r+s), (r·s), and r* are regular expressions denoting the languages $R \cup S$, $R \cdot S$, and R^*, respectively.

(3) Nothing else is a regular expression.

When no confusion is possible, some parentheses can be omitted. From here on, the language described by a regular expression r is denoted by L(r). Given a regular expression r, we use r^+ instead of $r \cdot r^*$. Obviously, $L(r^+) = L(r)^+$. The languages described by the regular expressions are regular, and vice versa.

THEOREM 3.1 If L is a regular language, then there is an NDFA accepting L.

Proof: by constructing inductively NDFA's with Λ-moves.

Suppose $M = (Q, I, \delta, q_0, F)$ is an NDFA. Note that $L(M)$ is the set of all assignments having paths from q_0 to some final states. Thus, we define for arbitrary $q, q' \in Q$ that

$$L_{qq'} = \{w \in I^* \mid \text{there is a path from } q \text{ to } q', \text{ whose assignment is } w\}.$$

That is, $L_{qq'}$ is the set of all assignments of the paths from q to q'. Clearly,

$$L(M) = \bigcup_{q \in F} L_{q_0 q}$$

Thus, if we can prove that for any $q, q' \in Q$, $L_{qq'}$ is regular, then $L(M)$ is regular.

THEOREM 3.2 The language accepted by an NDFA is regular.

Proof: assume that the state set of the NDFA is $\{1, 2, \ldots, n\}$. Following the above analyses, it suffices to prove that L_{ij}, $1 \leq i, j \leq n$, is regular.

For $0 \leq k \leq n$, define

$$L_{ij}^k = \{w \in I^* \mid \text{there is a path from } i \text{ to } j, \text{ whose assignment is } w,$$
$$\text{without going through any state higher than } k.\}$$

By 'going through a state', we mean both entering and then leaving.

We proceed to prove that L_{ij}^k is regular by induction on k.

BASIS: $k = 0$. L_{ij}^0 is the set of all assignments of the paths from i to j without passing through any state, i.e. the set of all assignments of the edges from i to j, therefore L_{ij}^0 is regular (note that if $i = j$, then $\Lambda \in L_{ij}^0$. If $i \neq j$, L_{ij}^0 is possibly an empty set, $\emptyset$).

INDUCTION: Suppose L_{ij}^{k-1} is regular for any $i, j \in \{1, 2, \ldots, n\}$. There are two cases for a path p from i to j without going through any state higher than k.

Case 1. The path never passes through a state as high as k. In this case, the assignments of the path are in L_{ij}^{k-1}; or

Case 2. The path p passes through the state k. In this case, p can be divided into three segments, $p = p_1 p_2 p_3$, where p_1 denotes the segment from i to k for the first time; p_2 denotes the segment from k back to k several times; p_3 denotes the segment from state k to j. It is clear that the assignment of p_1 is in L_{ik}^{k-1}, the assignment of p_2 is in $(L_{kk}^{k-1})^*$, and the assignment of p3 is in L_{kj}^{k-1}. Hence, the assignment of p is in $L_{ik}^{k-1} \cdot (L_{kk}^{k-1})^* \cdot L_{kj}^{k-1}$.

In either case, we have

$$L_{ij}^k = L_{ij}^{k-1} \cup L_{ik}^{k-1} \cdot (L_{kk}^{k-1})^* \cdot L_{kj}^{k-1} \qquad (3.1)$$

Therefore, for each $i, j \in Q$, L_{ij}^k is regular, which completes the induction.
Since $L_{ij} = L_{ij}^n$, we acquire that for each i,j, L_{ij} is regular as desired.
From Theorem 3.1 and Theorem 3.2, we have

KLEENE THEOREM. L is regular $\leftrightarrow$ there exists a finite automaton accepting L.

§4 PROPERTIES OF REGULAR LANGUAGES

The pumping lemma is an essential condition for a language to be regular, and is very useful.

THEOREM 4.1 (pumping lemma) Let I be an alphabet. For each regular language over the alphabet I, there exists a constant n such that if $w \in L$, and $|w| \geq n$, then there exist $x, y, z \in I^*$ satisfying

 (a) $w = xyz$;

 (b) $|xy| \leq n$, $|y| \geq 1$;

 (c) $xy^k z \in L$ $(k = 0,1,2,\ldots)$

Proof: because L is regular, there is a DFA accepting L. Let M be a DFA accepting L with n states. It should be clear that any path of length not less than n contains at least one cycle.

If $|w| \geq n$, then the path, p, corresponding to w is of length not less than n. Hence, there is at least one vertex, say q, that path p passes through twice. Thus, the path p may be divided into three segments: p_1 (from q_0 to q for the first time), p_2 (from q back to q), p_3 (the remainder).

Obviously,
$$|\bar{p}_1| + |p_2| \leq n, \quad |p_2| \geq 1.$$

Let p_1, p_2 and p_3 be labelled with x,y and z, respectively. Then

$$w = xyz, \quad |xy| \leq n, \quad |y| \geq 1.$$

Clearly, $p_1 p_2^k p_3$ (k = 0,1,2,...) and p have the same initial and final state, so $xy^k z$, the assignment of the path $p_1 p_2^k p_3$, is in L as desired.

By the definition of regular sets, the class of regular sets is closed under union, concatenation and Kleene closure. Now, some further properties of the regular sets can be shown.

THEOREM 4.2 Let L be a regular language over an alphabet I, then so is $L^c = I^* - L$.

Proof: suppose $M = (Q, I, \delta, q_0, F)$ is a DFA, and $L(M) = L$. Set $M' = (Q, I, \delta, q_0, Q-F)$. It is clear that $L(M') = L^c$. So, L^c is regular as desired.

THEOREM 4.3 Let both L_1 and L_2 be regular. Then, $L_1 \cap L_2$ and $L_1 - L_2$ are regular.

Proof: since

$$L_1 \cap L_2 = (L_1^c \cup L_2^c)^c,$$

and

$$L_1 - L_2 = L_1 \cap L_2^c,$$

this proposition is obviously true, by Theorem 4.2.

THEOREM 4.4 Let L be regular. $L^r = \{w^r | w \in L\}$ is then regular.

Proof: construct the reversed graph of the state transition diagram for M.

DEFINITION 4.1 Let I_1 and I_2 be two alphabets, and ϕ is a mapping from I_1^* to I_2^*. If ϕ satisfies the property:

$$\phi(xy) = \phi(x)\phi(y) \quad \text{for each } x,y \text{ in } I_1^*,$$

then, ϕ is called a homomorphism from I_1^* to I_2^*.

For any homomorphism ϕ, we have $\phi(\Lambda) = \Lambda$. In fact, $\phi(\Lambda) = \phi(\Lambda) \cdot \phi(\Lambda)$. Thus, $|\phi(\Lambda)| = 0$. Clearly, each homomorphism ϕ from I_1^* to I_2^* causes a mapping ψ from I_1 to I_2^* such that for any $a \in I_1$, $\psi(a) = \phi(a)$. Conversely, each mapping ψ from I_1 to I_2^* uniquely determines a homomorphism ϕ from I_1^* to I_2^* such that for any $a \in I_1$, $\phi(a) = \psi(a)$.

In fact, ϕ is defined by

$\phi(\Lambda) = \Lambda$

$\phi(wa) = \phi(w)\psi(a)$ for any $w \in I_1^*$, $a \in I_1$.

Therefore, $\phi(a_1 a_2 \ldots a_n) = \psi(a_1)\psi(a_2)\ldots\psi(a_n)$.

Let ϕ be a mapping from I_1^* to I_2^*, and $L_1 \subseteq I_1^*$, $L_2 \subseteq I_2^*$. Define $\phi(L_1) = \{\phi(w) | w \in L_1\}$, and $\phi^{-1}(L_2) = \{w \in I_1^* | \phi(w) \in L_2\}$. We have

<u>THEOREM 4.5</u> If ϕ is a homomorphism from I_1^* to I_2^*, and L_1 is a regular set over I_1, then $\phi(L_1)$ is a regular set over I_2.

Proof: by induction.

<u>THEOREM 4.6</u> Let ϕ be a homomorphism from I_1^* to I_2^*, and let L_2 be a regular set over I_2. Then $\phi^{-1}(L_2)$ is a regular set over I_1.

Proof: Let $M_2 = (Q, I_2, \delta_2, q_0, F)$ be a DFA accepting L_2. Define another DFA as follows:

$M_1 = (Q, I_1, \delta_1, q_0, F)$,

where $\delta_1(q,a) = \delta_2(q,\phi(a))$ (for all q in Q, and a in I_1).

It is easily shown that

$\delta_1(q,w) = \delta_2(q,\phi(w))$ (for all q in Q, and w in I_1^*).

Thus,

$\delta_1(q_0,w) \in F \leftrightarrow \delta_2(q_0,\delta(w)) \in F$,

namely, $w \in L(M_1) \leftrightarrow \phi(w) \in L(M_2) = L_2$.
So, $\phi^{-1}(L_2)$, being $L(M_1)$, is regular.

§5 TWO-WAY DETERMINISTIC FINITE AUTOMATA

For a DFA, the tape head must move one square to the right after reading one symbol. Now, the tape head is allowed the ability to move to the left as well as to the right, or remain stationary. Such a deterministic finite automaton is called a two-way deterministic finite automaton (2DFA). It begins with the tape head pointing at the rightmost symbol of the input string.

<u>DEFINITION 5.1</u> A two-way deterministic finite automaton (2DFA) is a quintuple $M = (Q,I,\delta,q_0,F)$, where Q,I,q_0, and F are as before, and δ is a mapping from $Q \times (I \cup \{\sqcup\})$ to $Q \times \{R,L,S\}$, called the next move function. Here, R, L and S denote moving the tape head one square to the right, one square to the left, and remaining stationary, respectively.

For an input string $a_1 a_2 \ldots a_n$ ($a_i \in I$), when M is in q, and its tape head is scanning the ith symbol a_i, we say that the instantaneous description (ID) of M at this time is

$$a_j \ldots a_{i-1} q a_i \ldots a_n \quad (j \leq 1,\ a_0 = a_{-1} = \ldots \sqcup).$$

Notice that the ID $a_1 \ldots a_n q$ indicates that the tape head falls off the right end of the input with the FC in state q, and that the case $i \leq 0$ denotes that the tape head falls off the left end of the input. The blanks in the left end can be deleted. No matter how many blanks are added to the left it will make no difference.

Suppose at some time the ID of M is $a_j \ldots a_{i-1} q a_i \ldots a_n$.

If $i \leq n$, and $\delta(q,a_i) = (q',R)$, then the next ID of M is $a_j \ldots a_i q' a_{i+1} \ldots a_n$. (Note that what it implies if $i = n$.)

If $\delta(q,a_i) = (q',S)$, then we say the next ID of M is $a_j \ldots a_{i-1} q' a_i \ldots a_n$.

If $\delta(q,a_i) = (q',L)$, then the next ID of M is $a_j \ldots a_{i-2} q' a_{i-1} \ldots a_n$. (Note that sometimes a blank needs to be added to left.)

It is termed one move of M when M goes from one ID to another by the above rules. If M goes from ID_1 to ID_2 in one move, we write $ID_1 \vdash ID_2$; if M enters ID_2 from ID_1 by a finite number of moves, including zero moves, we write $ID_1 \vdash^* ID_2$.

Finally, define the language accepted by M as follows:

$$L(M) = \{a_1a_2...a_n \in I^* | \text{ for some } q \text{ in } F, a_1a_2...q_0a_n \vdash^* a_1a_2...a_nq\}$$

Now it can be proved that the class of languages accepted by 2DFA's is also regular. The idea behind the proof is as follows: for each string in I^+, we assume that initially the 2DFA is in state q, its tape head scanning the rightmost symbol, and after a finite number of moves the head will eventually move off the right end of the input and enter state q'. Then, clearly, q' is uniquely determined by w and q, that is, w causes a mapping from Q to Q:

$$\sigma_w : q \to q'.$$

Note that since the tape head may never fall off the right end of w for some q's, the domain of this mapping is some subset of Q rather than Q itself. Such a mapping is called a partial mapping on Q. It is not difficult to see that σ_{wa} is uniquely determined by σ_w and a, so δ' is a mapping, where

$$\delta' : (\sigma_w, a) \to \sigma_{wa}.$$

This implies that a DFA can be made to simulate the previous 2DFA by using the set $\{\sigma_w | w \in I^*\}$ as a new state set. Let $M = (Q, I, \delta, q_0, F)$ be a 2DFA. For each string $w = a_1a_2...a_n$, define the partial mapping from Q to Q as follows:

$$\sigma_w(q) = \begin{cases} q', & \text{if } a_1a_2...qa_n \vdash^* a_1a_2...a_nq'; \\ \text{undefined}, & \text{otherwise.} \end{cases}$$

Define σ_Λ as the transition from Q to Q caused by the tape head falling off the left end and moving back.

Define

$$M' = (Q', I, \delta', \sigma_\Lambda, F'),$$

where

$$Q' = \{\sigma_w | w \in I^*\},$$
$$\delta'(\sigma_w, a) = \sigma_{wa},$$
$$F' = \{\sigma_w | \sigma_w(q_0) \in F\}.$$

Thus, M' is a DFA, and has no more than $(|Q| + 1)^{|Q|}$ states. It is easy to prove that $\delta'(\sigma_\Lambda, w) = \sigma_w$.

Furthermore, for $w = a_1 a_2 \ldots a_n$,

$$w \in L(M') \leftrightarrow \delta'(\sigma_\Lambda, w) \in F' \leftrightarrow \sigma_w \in F' \leftrightarrow \sigma_w(q_0) \in F$$

$$\leftrightarrow a_1 a_2 \ldots q_0 a_n \vdash^* a_1 a_2 \ldots a_n q, \text{ for some } q \text{ in } F \leftrightarrow w \in L(M).$$

So, we have

THEOREM 5.1 For each 2DFA with k states, there exists an equivalent DFA with at most $(k+1)^k$ states.

In other words, 2DFA and DFA can simulate each other, i.e. 2DFA $\leftrightarrow$ DFA. Thus, in this chapter we have shown that

$$\text{2DFA} \leftrightarrow \text{DFA} \leftrightarrow \text{NDFA} \leftrightarrow \text{R.E.}$$

2 The Turing machine

§6 COMPUTABILITY

What is computation? This question seems so simple that any pupil in a primary school could answer: the four arithmetic operations are computation! However, modern computers, the machinery of computation, are used not only in scientific calculations, but also in various other fields, such as management, translation, theorem proving, prediction, and so on. They can even talk with their users and do housework. Such computers are called robots. After all, what is computation? What problem is computable? Is there any problem which is not computable? These questions are not only important in mathematics, but also of great importance to science and philosophy.

As early as 1936, A. Turing made a penetrating analysis of computation. The usual computation, he thought, is performed by a man with a pen and a piece of paper. The man erases or writes some symbols on the paper according to the symbols he is looking at and some rules stored in his brain, then he moves his eyes to other places, repeats the above process until he believes that the computation is complete. The result of the computation is the symbols written at specific positions on the paper.

Informally, the human brain can be regarded as a finite state control (FC), the paper as a tape, and the pen as a read/write head. The tape is divided into infinite squares, each square holding one symbol from a given finite alphabet. According to the symbol under the read/write head (i.e. the symbol the man is looking at) and the current state of FC (i.e. the state in his brain), the FC (i.e. the brain of the man) can make the following three decisions:

 (a) Changes its current state (for the next operation),
 (b) Changes the symbol scanned by the read/write head (printing a symbol to replace what was there), and
 (c) Moves the read/write head left or right one square, or keeps it stationary.

Thus, Turing constructed a very simple but quite satisfactory model for the computing procedure (for details see the next section). Such a mechanism is called a Turing machine.

Dividing the computing procedure into primitive, simple and mechanical steps, Turing succeeded in constructing a computational model. It appears that this model is too simple. But, as Turing discovered, it can perform quite difficult computations. Thus, Turing declared that a problem is computable if his machine can compute it; otherwise, the problem is not computable.

This appears too subjective. Imagine the controversy if someone else designed a computational model and made the same declaration as Turing. However, in the 50 years since the publication of Turing's paper, scientists have done their best to design various but reasonable models of computation, and have found that all reasonable models are equivalent to the Turing machine in computing power. This reflects the well-known Church-Turing thesis which states that any reasonable computational models are equivalent to each other.

Of course, this thesis cannot be proved mathematically, since there is no way to define what a reasonable computational model is. It can only be proved in a historical course. The fact that any reasonable computational model is equivalent to the Turing machine explains that computability itself is independent of any specific reasonable model, which is the foundation of modern theoretical computer science.

§7 TURING MACHINE

As mentioned in the preceding section, a Turing machine is composed of a finite state control (FC), a read/write tape head and a tape whose both ends are unbounded (Fig. 7.1).

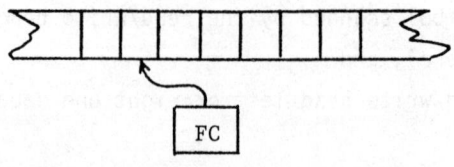

Fig. 7.1

DEFINITION 7.1 A Turing machine (TM) is a 7-tuple

$$M = (Q, I, \Sigma, \delta, \sqcup, q_0, F),$$

where,

Q is the finite set of states,
Σ is the finite set of allowable tape symbols, called the work alphabet,
$I \subset \Sigma$ is the set of input alphabet,
$\sqcup$ in $\Sigma - I$ is the blank,
q_0 in Q is the initial state,
$F \subseteq Q$ is the set of final states,
δ is the next move function, a mapping from a subset of $Q \times \Sigma$ to $Q \times \Sigma \times \{R, L, S\}$.

Note that $\delta(q, a) = (q', a', d)$ means that the machine, in state q and scanning the symbol a, will enter state q', replace a by a' and move the tape head in the direction of d.

Given an input word w over I, first put w on the tape, let the head scan the leftmost symbol of w and put the machine in state q_0. Then, by the rules of δ, M moves stepwise until at some step δ is undefined, or M enters some state q in F. At this point, the machine halts. If M halts in some state q in F, then we say w is accpeted by M, and call the current contents of the tape the computed result $M(w)$; otherwise, w is rejected by M.

An instantaneous description (ID) of M is a string in $\Sigma^* Q \Sigma^*$. The ID xqy means that M is in state q with contents xy on the tape, and the tape head is scanning the leftmost symbol of y (if $y = \Lambda$, the head is scanning a blank).

Between two ID's we define the relation $\vdash$ as follows: if $q \notin F$, then

$$a_1 \ldots a_{i-1} q a_i \ldots a_m \vdash \begin{cases} a_1 \ldots a_{i-1} q' a' a_{i+1} \ldots a_m & \text{(if } \delta(q, a_i) = (q', a', S)), \\ a_1 \ldots a_{i-2} q' a_{i-1} a' a_{i+1} \ldots a_m & \text{(if } \delta(q, a_i) = (q', a', L)), \\ a_1 \ldots a_{i-1} a' q' a_{i+1} \ldots a_m & \text{(if } \delta(q, a_i) = (q', a', R)). \end{cases}$$

Note that sometimes a blank symbol should be added to the end of the description word. The relation $\vdash$ represents a move of M; that is, if D_1, D_2 are two ID's of M, then $D_1 \vdash D_2$ represents that the machine can go to D_2 from D_1 in one move. The closure of the relation $\vdash$ is denoted by $\vdash^*$. $D_1 \vdash^* D_2$ means M can arrive at D_2 from D_1 by a sequence of moves, including zero moves.

Given $w \in I^*$, if there exists some state q in F such that $q_0 w \vdash^* xqy$, then w is said to be accepted by M, and M(w) is defined as M(w) = xy; otherwise, M(w) is undefined and w is rejected by M. So, for w rejected, it is possible that the machine M either will never halt, or halt in a non-accepting state.

The language accepted by M is

$L(M) = \{w \in I^* | M \text{ accepts } w\}$.

Thus, we have strictly defined M(w) (for all w in I*) and L(M). In effect, two functions of M have been mentioned; they are

(a) Language acceptor: the language accepted is L(M).
(b) Sequence transducer: T realises the transformations $w \mapsto M(w)$, for all w in I^*.

§8 MULTITAPE TURING MACHINES

Informally, a multitape Turing machine is composed of a finite state control (FC), k tapes and k tape heads. Each tape is the same as in a single-tape Turing machine. But all the k tape heads are controlled by one FC. According to the symbols under the heads (k symbols) and its own current state, the FC does the following three jobs:

(1) changes states;
(2) prints a new symbol on each of the squares under its tape heads;
(3) moves each of its tape heads, independently, one square to the left or right, or keeps them stationary.

DEFINITION 8.1 A k-tape Turing machine is a 7-tuple

$M = (Q, I, \Sigma, \delta, \sqcup, q_0, F)$

where $Q, I, \Sigma, \sqcup, q_0, F$ are as in Definition 7.1; δ, the next move function, is a mapping from a subset of $Q \times \Sigma^k$ to $Q \times \Sigma^k \times \{L,R,S\}^k$.

It is stipulated that some tapes are input tapes and some are output tapes (input and output tapes may overlap).

Initially, the i input tapes store i words over I, $w_1, w_2, \ldots, w_i$ (each tape stores one word), each input-tape head points to the leftmost symbol of its tape content, and all the other tapes are blank. The FC is in the initial state q_0. Then, by the rules of δ, M moves stepwise. If M finally halts with its FC entering some state in F, we say that $M(x_1,x_2,\ldots,x_i)$ is defined. Suppose that the immediate contents of the j-output tapes are $y_1,y_2,\ldots,y_j$. Then define $M(x_1,x_2,\ldots,x_i) = (y_1,y_2,\ldots,y_j)$. Otherwise $M(x_1,x_2,\ldots,x_i)$ is undefined.

Note that no formal definition has been given of the moves of M. However, after the fashion of §7, the reader can write out it by himself. It should be noted that in such case an ID of M is a k-tuple

$$(w_1qu_1, w_2qu_2, \ldots, w_kqu_k),$$

where q is the FC's state, $w_\ell u_\ell$ is the contents of the ℓth tape, and the ℓth tape head is scanning the leftmost symbol of u_ℓ.

As a language accepter, a multitape Turing machine should be viewed as a special version with only one input tape and no output tape. For the input $w \in I^*$, the machine determines whether w is accepted depending on whether M halts in a state in F.

For any k-tape Turing machine M_1, there exists a single-tape Turing machine M_2 simulating M_1. By "simulating", we mean that if the input alphabet and output alphabet of M_1 are I and Σ respectively, then the input alphabet of M_2 is $I \cup \{\Delta\}$, where $\Delta \notin \Sigma$ (it is a separator), and

$$M_1(x_1,\ldots,x_i) = (y_1,\ldots,y_j) \leftrightarrow M_2(x_1\Delta\ldots\Delta x_i) = y_1\Delta\ldots\Delta y_j.$$

Note that if either end is defined, then the other is also defined.

We can first design a single type TM T_2' as follows. On its tape there are 2k tracks, i.e. we can put 2k work symbols of T_1 into one square of the tape of T_2' (in other words, the work alphabet of T_2' is Σ^{2k}, where Σ is the work alphabet of T_1). Track i is used to store the contents of tape i of T_1 (i = 1,2,...,k). Track i+k is used to mark the position of the i-th tape head (i = 1,2,...,k).

Initially, the input words of T_1 are in the corresponding tracks. In order to simulate one move of T_1, the machine T_2' moves its tape head from the left end to the right end of its contents to find all the k symbols scanned by T_1. It can remember these symbols and the state of T_1 by its inner states. After knowing all information needed to simulate this move, the machine T_2' then moves its head from the right end to the left end of its contents and completes the simulation of one move.

Now it is not difficult to design a single tape TM T_2, which first transforms the input $x_1 \Delta x_2 \Delta \ldots \Delta x_i$ to the input form of T_2' then simulates T_2', and finally transforms the output of T_2' to the form $y_1 \Delta y_2 \Delta \ldots \Delta y_j$.

Thus, we have

THEOREM 8.1 For any multitape Turing machine, there exists a single-tape Turing machine simulating it.

§9 TURING MACHINE COMBINATION

A multitape Turing machine M can be regarded as a transform which converts the inputs to the outputs. Suppose M's input tapes are $X_1, X_2, \ldots, X_i$, and the output tapes, $Y_1, Y_2, \ldots, Y_j$. Then this transform can be written as

$$(Y_1, \ldots, Y_j) := M(X_1, \ldots, X_i).$$

If $w_1, \ldots, w_j$ are j words, then

$$M(X_1, \ldots, X_i) = (w_1, \ldots, w_j)$$

is a predicate (a condition). Given an input $x_1, \ldots, x_i$ the value of the predicate is possible to be TRUE or FALSE. We can construct a Turing machine to test the predicate. The Turing machine has two final states, TRUE and FALSE. First, it copies the contents of $X_1, \ldots, X_i$ onto somewhere, then simulates the behaviour of M obtaining j results on the tapes $Y_1, \ldots, Y_j$ respectively, and then compares the j results with $w_1, \ldots, w_j$; if they are the same, then the Turing machine enters state TRUE, otherwise it enters state FALSE, but before entering the final states, it regains the contents of $X_1, \ldots, X_i$ and erases all the other tapes. If this TM halts, then the values of the input tapes remain the same, the other tapes are all blank, and the TM enters state TRUE iff the predicate is true.

In the following three basic kinds of combinations of TM are presented.

1. <u>Composition</u>

Let M_1 and M_2 be two Turing machines realizing the transforms M_1 and M_2 respectively. Given an input, M_1 produces a corresponding output. Taking this output as the input to M_2, we get M_2's output for the given input. Thus, a new TM M is gained, which realizes the composition of M_1 and M_2. M is called the composition of M_1 and M_2, denoted by $M = M_1 \cdot M_2$, or "$M_1 ; M_2$".

2. <u>Branch</u>

Let C be a condition, and M_1 and M_2 two Turing machines. First, we construct a TM that tests the condition C. For an input, if this TM enters state TRUE, then M_1 works on the input and produces a corresponding output; if state FALSE is entered, then M_2 works on the input and also produces a corresponding output. Thus, we obtain a new TM M which realizes the branching under condition C. This TM is denoted as

   IF C THEN M_1 ELSE M_2.

3. <u>Loop</u>

Let C be a condition, and T a TM. A new TM is constructed as follows: first, test condition C, if it is TRUE, then let T work on the input and get an output; then, take the output as an input and test condition C again,..., until the state entered is FALSE, then take the final result of T as the output. Such a TM is denoted as

   WHILE C DO T.

§10 THE UNIVERSAL TURING MACHINE

Each Turing machine described so far is for a "special purpose", which can only realize one particular transformation. But it is not so for a modern computer. In fact, as long as various programs are stored in the memory, the machine can accomplish various tasks. So, can a universal Turing machine (U) be designed such that if appropriate programs are stored on its "program tape", it can simulate any particular Turing machine M?

What does it mean for U to simulate M? Simply, for the same input they produce the same output. However, the alphabets of Turing machines may be of various kinds, and the alphabet of U is finite. How can U simulate any Turing machine M? There is a simple way, i.e. for each alphabet Σ to design a universal Turing machine.

But another method. That is, set the alphabet $\Sigma = \{0, 1, *, \times, \sqcup\}$. For any arbitrary Turing machine M, encode M's input with 0-1 code ($*, \times$ serve as separators). The result of U on this coding should be the coding of M's result. This is an intuitive meaning for U to simulate M.

Since the single-tape Turing machines have the same power as the multitape Turing machines in computing, without loss of generality, we design a multitape Turing machine U with work alphabet $\Sigma = \{0,1,*,X,\sqcup\}$, such that for any arbitrary single-tape Turing machine M there is a descriptor $d_M \in \{0,1,*,X\}*$ so that for each input word w,

$$M(w) \text{ is defined} \leftrightarrow U(C(w),d_M) \text{ is defined, and}$$

$$C(M(w)) = U(C(w),d_M), \text{ if } M(w) \text{ is defined.}$$

(10.1)

where C(w) denotes the coding of w.

Suppose

$$M = (Q,I,\Sigma,\delta,q_0,\{q_f\})$$

is a single-tape Turing machine with a single final state q_f with the tape head moving either left (L) or right (R). Assume

$$|\Sigma| \leq 2^k, \quad |Q| \leq 2^n.$$

Thus,

(1) Each symbol $a \in \Sigma$ can be represented by C(a), a binary string of length k, which is called a's coding. The coding of a word over Σ is defined as follows:

$$C(\Lambda) = \Lambda$$

$$C(wa) = C(w)C(a) \text{ for all } w \text{ in } \Sigma^*, a \text{ in } \Sigma.$$

(2) Each state $q \in Q$ can be represented by C(q), a binary string of length n, called the coding of q. And set

$$C(q_0) = 0...0 = 0^n$$

$$C(q_f) = 1...1 = 1^n.$$

(3) The function δ is encoded as follows: for "$\delta(q,a) = (q',a',d)$", its coding is

$$XC(q)*C(a)* C(q')*C(a')*C(d)$$

where $C(L) = 1$, $C(R) = 0$. While δ's coding, $C(\delta)$, is the concatenation of all the possible codings of the above form. Let

$$d_M = C(\delta)X,$$

be the descriptor of M.

The universal Turing machine U has four work tapes, the 1st and the 4th are the input tapes (the 4th tape is used to store d_M), also the 1st tape is the output tape.

U simulates M as follows:

Suppose d_M is stored on the 4th tape and on the 1st one is the coding of M's input word w, $C(w)$. The 3rd tape will be used to store the current state of M and the symbol scanned by M. The 2nd tape will be used to store the length k.

First, U is initialized: write 0^k on the 2nd tape, i.e. give the length of codings of symbols in Σ. On the 3rd tape write 0^n*, denoting M in state q_0. Move each tape head to the leftmost symbol of its tape contents. Note that the numbers k and n can be obtained from d_M.

Then, U simulates each move of M by the following five steps:

STEP 1 READ-SYMBOL: copy the contents of k squares right to the head of the 1st tape onto the 3rd tape right to * (k can be determined by the contents of the 2nd tape), then the 1st, 2nd and 3rd tape heads, respectively, move back to the original positions before this step.

STEP 2 LOOK-FOR-TABLE: suppose that the contents of the 3rd tape are $C(q)*C(a)$; on the 4th tape look for the 3rd tape's contents after each X.

If the 3rd tape's contents cannot be found on the 4th tape, then halt with U not entering the final state.

If found, then if the contents of this segment on the 4th tape are

$$XC(q)*C(a)*C(q')*C(a')*C(d)X,$$

move the 4th tape head to the leftmost symbol of $C(q')$, and erase completely the 3rd tape.

STEP 3 CHANGE-STATE: write down $C(q')*$ on the 3rd tape, move the 3rd tape head to the leftmost symbol of the contents, and move the 4th tape head to the leftmost symbol of $C(a')$.

STEP 4 PRINT-SYMBOL: replace the symbols on the k squares to the right of the 1st head by C(a'), move the 1st, 2nd heads back to their original positions respectively, and move the 4th head to C(d).

STEP 5 MOVE-HEAD: move the 1st head in the direction indicated by C(d) (left or right) k squares, and move the 2nd and 4th heads to the leftmost symbols of their contents respectively.

Finally, check the condition NOT-HALT, i.e. the contents of the 3rd tape are not 1^n* (corresponding to state q_f). If the condition NOT-HALT holds, then go back to STEP 1, otherwise enter the final state.

So, U can be described as

BEGIN
  INITIAL;
  WHILE NOT-HALT DO
  (READ-SYMBOL; LOOK-FOR-TABLE; CHANGE-STATE; PRINT-SYMBOL; MOVE-HEAD)
END

It should be clear that U simulates M in the sense of (10.1). So, we have

THEOREM 10.1 The universal Turing machine U exists in the sense of (10.1).

3 Recursive functions

§11 DEFINITIONS AND EXAMPLES

Some basic functions are presented below:

(a) The zero function $O(x) = 0$ (for all x in N);

(b) The successor function $S(x) = x'$ (for all x in N), where $x' = x+1$ is the successor of x;

(c) The projection function $I_i^n(x_1, x_2, \ldots, x_n) = x_i$ (for all $x_1, x_2, \ldots, x_n$ in N).

We now introduce a method of constructing new functions.

DEFINITION 11.1 (PRIMITIVE RECURSION) Suppose that $\psi(x_1, \ldots, x_n)$, $\phi(y, x_1, \ldots, x_n)$ and $\chi(z, y, x_1, \ldots, x_n)$ are functions of n-variables, (n+1)-variables and (n+2)-variables on N, respectively, and satisfy the following relations:

$$\phi(0, x_1, \ldots, x_n) = \psi(x_1, \ldots, x_n),$$

$$\phi(y', x_1, \ldots, x_n) = \chi(\phi(y, x_1, \ldots, x_n), y, x_1, \ldots, x_n),$$

for all $x_1, \ldots, x_n, y$ in N. Then, we call ϕ primitive recursive in ψ and χ.

DEFINITION 11.2 (a) The functions $O(x)$, $S(x)$ and $I_i^n(x_1, \ldots, x_n)$ ($n = 1, 2, 3, \ldots$; $i = 1, 2, \ldots, n$) are all primitive recursive functions.

(b) If ψ and χ are both primitive recursive functions, and ϕ is primitive recursive in ψ and χ, then ϕ is a primitive recursive function.

(c) If $\psi(x_1, \ldots, x_n)$ and $\chi_1(y_1, \ldots, y_m)$, $\chi_2(y_1, \ldots, y_m), \ldots, \chi_n(y_1, \ldots, y_m)$ are all primitive recursive functions, then so is the function

$$\phi(y_1, \ldots, y_m) = \psi(\chi_1(y_1, \ldots, y_m), \chi_2(y_1, \ldots, y_m), \ldots, \chi_n(y_1, \ldots, y_m)).$$

(d) Nothing else is a primitive recursive function.

EXAMPLE 11.1 Suppose $f(x)$ is a primitive recursive function (p.r.f.) and $g(x,y) = f(x)$. Then $g(x,y)$ is a p.r.f.

Proof: $g(x,y) = f(I_1^2(x,y))$.

Similarly, an n-ary p.r.f. is also p.r.f. when viewed as a (n+1)-ary function.

EXAMPLE 11.2 The constant function $C_k(x) = k$ (for all x in N) is a p.r.f.

Proof: induction on k.

EXAMPLE 11.3 Let $f(x,y)$ be a p.r.f, k a natural number, and $g(x) = f(x,k)$. Then, $g(x)$ is a p.r.f.

Proof: since $g(x) = f(I_1^1(x), C_k(x))$, and f, I_1^1, and C_k are p.r.f., so $g(x)$ is a p.r.f.

EXAMPLE 11.4 $f^+(x,y) = x+y$ is a p.r.f.

Proof: $f^+(0,y) = I_1^1(y)$,

$f^+(x',y) = S(f^+(x,y))$.

EXAMPLE 11.5 $f^*(x,y) = xy$ is a p.r.f.

Proof: $f^*(0,y) = 0(y)$,

$f^*(x',y) = f^+(f^*(x,y),y)$.

Note that the recursive equations in Example 11.4 can be written as

$0 + y = y$,

$x' + y = S(x+y)$;

This written form will be used here after.

EXAMPLE 11.6 $x \dotdiv y$ is a p.r.f, where $x \dotdiv y =$ if $x \geq y$ then $x-y$ else 0.

Proof: first, $0 \dot{-} 1 = 0$

$$x' \dot{-} 1 = x.$$

So, $x \dot{-} 1$ is a p.r.f. Second,

$$x \dot{-} 0 = x$$
$$x \dot{-} y' = (x \dot{-} y) \dot{-} 1,$$

so, $x \dot{-} y$ is a p.r.f.

EXAMPLE 11.7 $|x-y|$ is a p.r.f.

Proof: $|x-y| = (x \dot{-} y) + (y \dot{-} x)$.

EXAMPLE 11.8 If $f(x,y)$ is a p.r.f, then so are $\sum_{x=0}^{y} f(x,y)$ and $\prod_{x=0}^{y} f(x,y)$.

Proof: let

$$g(0,y) = f(0,y),$$

$$g(z',y) = g(z,y) + f(S(z),y).$$

Obviously,

$$g(z,y) = \sum_{x=0}^{z} f(x,y)$$

is a p.r.f. Thus,

$$g(I_1^1(y), I_1^1(y)) = g(y,y) = \sum_{x=0}^{y} f(x,y)$$

is a p.r.f.

EXAMPLE 11.9 $\chi_0(x)$ is a p.r.f, where $\chi_0(x) = $ if $x = 0$ then 1 else 0.

Proof: $\chi_0(0) = 1$,
$\chi_0(x') = 0$.

EXAMPLE 11.10 Let $\ell \neq 0$. Then, $q_\ell(x) = \lfloor x/\ell \rfloor$, and $r_\ell(x) = x - \ell \cdot \lfloor x/\ell \rfloor$ are both p.r.f. (Note that $q_\ell(x)$ is the quotient of x divided by ℓ, and $r_\ell(x)$ is the remainder.)

Proof: since

$$q_\ell(x) = \sum_{y=1}^{x} X_o(\ell \cdot y \dot{-} x)$$

so, $q_\ell(x)$ is a p.r.f. Thus so does $r_\ell(x) = x \dot{-} \ell \cdot q_\ell(x)$.

EXAMPLE 11.11 The characteristic function of k, $X_k(x) = $ if $x = k$ then 1 else 0.

Proof: observe that $X_k(x) = X_o(|x-k|)$.

EXAMPLE 11.12 Let f(x) be a unary p.r.f. If we change the values of f(x) at a fixed number of points, then the function obtained is also a p.r.f. Particularly, a unary function that keeps constant except at a finite number of points is a p.r.f.

Proof: suppose the values of f(x) at $k_1, k_2, \ldots, k_r$ are changed into $C_1, C_2, \ldots, C_r$, respectively, and the function obtained is denoted by g(x).
  Obviously,

$$g(x) = f(x) + \sum_{i=1}^{r} (C_i - f(k_i)) X_{k_i}(x).$$

So, g(x) is a p.r.f.

Note that Examples 11.11 and 11.12 can be extended to the n-ary functions.

EXAMPLE 11.13 The function

$$E(x,y) = \text{if } (x=y) \text{ then } 1 \text{ else } 0$$

is a p.r.f.

Proof: $E(x,y) = X_o(|x-y|)$.

EXAMPLE 11.14 A one-to-one correspondence between $N \times N$ and N is represented by

$$Z(x,y) = y + (x+y)(x+y+1)/2.$$

Let the inverse of the above correspondence be

$$z \to (X(z), Y(z)).$$

Then $X(z)$, $Y(z)$, and $Z(x,y)$ are all p.r.f.

Proof: it should be clear that $Z(x,y)$ is a p.r.f. In addition, since

$$X(z) = \sum_{u=0}^{z} \sum_{v=0}^{z} u \cdot E(Z(u,v),z)$$

and

$$Y(z) = \sum_{u=0}^{z} \sum_{v=0}^{z} v \cdot E(Z(u,v),z),$$

where E is defined as in Example 11.13, so, $X(z)$ and $Y(z)$ are p.r.f.

EXAMPLE 11.15 Suppose $\phi(x,y,z)$, $\psi(x,y,z)$ are p.r.f., and $f(x)$, $g(x)$ satisfy the equations

$$f(0) = g(0) = 0$$

$$f(x') = \phi(x,f(x),g(x))$$

$$g(x') = \psi(x,f(x),g(x)). \qquad (11.1)$$

Prove that $f(x)$ and $g(x)$ are p.r.f.

Proof: let $h(x) = Z(f(x),g(x))$. Then,

$$f(x) = X(h(x)), \quad g(x) = Y(h(x)).$$

On the other hand,

$$h(0) = Z(f(0),g(0)) = Z(0,0) = 0$$

$$h(x') = Z(f(x'),g(x'))$$

$$= Z(\phi(x,X(h(x)),Y(h(x))),\psi(x,X(h(x)),Y(h(x)))).$$

Thus, $h(x)$ is a p.r.f. Therefore, $f(x)$ and $g(x)$ are p.r.f.

Note that this example can be viewed as a primitive recursion of "vector functions" and can be extended to n-dimension vector functions of m-variables.

Notice that a p.r.f. is always defined everywhere. Later, the functions defined everywhere are called total functions; the others, for emphasis, are called partial functions. The identity of two functions means that if one is defined at some point, then the other is also defined at this point, and the values are the same.

DEFINITION 11.3 (MINIMALISATION) Let $g(t,x_1,\ldots,x_n)$ be a total function. Define the function $f(x)$ as follows:

$$f(x_1,\ldots,x_n) = \mu_t(g(t,x_1,\ldots,x_n) = 0)$$

$$= \min\{t \mid g(t,x_1,\ldots,x_n) = 0\}.$$

We say that the function $f(x_1,\ldots,x_n)$ is obtained from $g(t,x_1,\ldots,x_n)$ using minimalisation on t.

Clearly, $f(x_1,\ldots,x_n)$ may not be a total function.

DEFINITION 11.4 (a) $O(x), S(x)$ and $I_i^n(x_1,\ldots,x_n)$ ($n = 1,2,3,\ldots$; $i = 1,2,\ldots,n$) are all recursive functions (r.f.).
  (b) If ψ, χ are both r.f., and ϕ is primitive recursive in ψ and χ, then ϕ is an r.f.
  (c) If $\psi(x_1,\ldots,x_n)$, $\chi_1(y_1,\ldots,y_m),\ldots,\chi_n(y_1,\ldots,y_m)$ are all r.f. then so is the function $\psi(\chi_1(y_1,\ldots,y_m),\ldots,\chi_n(y_1,\ldots,y_m))$.
  (d) If $\phi(t,x_1,\ldots,x_n)$ is a total r.f. then $\mu_t(\phi(t,x_1,\ldots,x_n) = 0)$ is an r.f.
  (e) Nothing else is a recursive function.

§12 THE ARITHMETIZATION OF TURING MACHINES

In this section it will be proved that each function computable by a Turing machine is recursive.

Let M be a single-tape Turing machine with only one final state, and suppose that only when M enters the final state does it halt. Moreover, the function δ is defined on $Q \times \Sigma$, the initial state and the final state are distinct. Whenever M halts, the read/write head is pointing to the leftmost symbol of the contents of the input tape. Before this, the read/write head can only move left or right one square in a move, but not remain stationary.

Note that this restriction does not decrease the computing power of a Turing machine. Also, we can assume that the machine has a special state such that the machine will never halt as long as it enters the state. Assume that the work alphabet Σ of M has ℓ elements, say, without loss of generality,

$$\Sigma = \{0,1,\ldots,\ell-1\},$$

where 0 denotes the blank. Thus, given any input string $w = a_0 a_1 \ldots a_n$, we can represent it by the natural number

$$C(w) = a_0 \ell^0 + a_1 \ell^1 + \ldots + a_n \ell^n,$$

where $x = C(w)$ is called the coding of the string w. If the corresponding output is $u = b_0 b_1 \ldots b_m$, and its coding is

$$y = C(u) = b_0 \ell^0 + b_1 \ell^1 + \ldots + b_m \ell^m,$$

then we say that M realizes a transform $x \to y$. In other words, the function computed by M is

$$f(x) = C(M(C^{-1}(x))) \quad (x \in N),$$

where $C^{-1}(x)$ denotes the string whose coding is x. Let the state set of M be

$$Q = \{0,1,\ldots,k\},$$

where 0 is the initial state, 1 is the final state.

Suppose, for the input string $w = C^{-1}(x)$ $(x \in N)$, M enters state $q(t,x)$ after t moves, and the present configuration of the tape is shown in Fig. 12.1. This configuration can be represented by two natural numbers m and n, where

$$m(t,x) = \sum_{0}^{\infty} b_k \ell^k, \quad n(t,x) = \sum_{0}^{\infty} c_k \ell^k$$

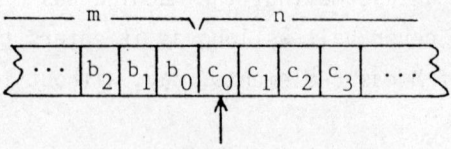

Fig. 12.1

Thus, the ID at moment t can be stated as

$(q,m,n) = (q(t,x), m(t,x), n(t,x))$.

At this moment the scanned symbol a is the 'lowest bit' of n, i.e. the remainder of n divided by ℓ:

$a = a(t,x) = r_\ell(n(t,x))$.

Let the next move function δ of M be

$\delta(q,a) = (Q(q,a), W(q,a), D(q,a))$ $(0 \leq q \leq k,\ 0 \leq a \leq \ell-1)$

That is, $Q(q,a)$ is the next state, $W(q,a)$ is the rewritten symbol, and $D(q,a)$ is the moving direction (0 for moving to the right, 1 for moving to the left). Notice that the functions Q, W and D are only defined at some finite points. At other points we extend the definition as follows:

$Q(q,a) = 2$

$W(q,a) = 0$

$D(q,a) = 0$ (if $\delta(q,a)$ is not defined)

where 2 denotes a special state; upon entering this state, the machine will never halt.

So Q, W and D are all p.r.f. (refer to Example 11.12).

Using the ID of M at moment t, $(q(t,x), m(t,x), n(t,x))$, we can compute the ID of M at moment (t+1):

34

$$q(t+1,x) = Q(q(t,x), r_\ell(n(t,x))).$$

If $D(q(t,x), r_\ell(n(t,x))) = 0$ (moving right) (see Fig. 12.2) then we have

$$m(t+1,x) = \ell \cdot m(t,x) + W(q(t,x), r_\ell(n(t,x)))$$

$$n(t+1,x) = q_\ell(n(t,x))$$

where q_ℓ, r_ℓ are the quotient and the remainder of n divided by ℓ, respectively (refer to Example 11.10).

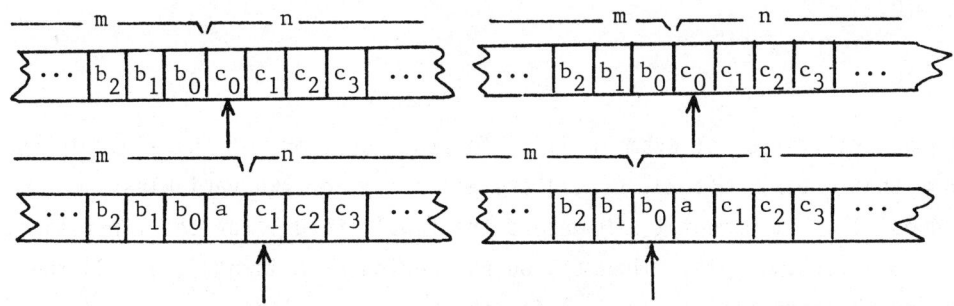

Fig. 12.2   Fig. 12.3

If $D(q(t,x), r_\ell(n(t,x))) = 1$ (moving left) (see Fig. 12.3) then we have

$$m(t+1,x) = q_\ell(m(t,x))$$

$$n(t+1,x) = \ell^2 \cdot q_\ell(n(t,x)) + \ell \cdot W(q(t,x), r_\ell(n(t,x))) + r_\ell(m(t,x)).$$

In summary, we have

$$q(t+1,x) = Q(q, r_\ell(n))$$

$$m(t+1,x) = \chi_0(D(q,r_\ell(n))) \cdot (\ell \cdot m + W(q,r_\ell(n))) + \chi_1(D(q,r_\ell(n))) \cdot q_\ell(m)$$

$$n(t+1,x) = \chi_0(D(q,r_\ell(n))) \cdot q_\ell(n) + \chi_1(D(q,r_\ell(n))) \cdot (\ell^2 q_\ell(n) + \ell \cdot W(q,r_\ell(n)) + r_\ell(m)),$$

where q, m and n in the right end denote $q(t,x)$, $m(t,x)$ and $n(t,x)$, respectively.

χ_0 and χ_1 denote the characteristic functions of 0 and 1. (Refer to Example 11.11.) Or, we have

$$q(t+1,x) = f_1(q(t,x),m(t,x),n(t,x))$$

$$m(t+1,x) = f_2(q(t,x),m(t,x),n(t,x))$$

$$n(t+1,x) = f_3(q(t,x),m(t,x),n(t,x))$$

where f_1, f_2 and f_3 are all p.r.f. of three variables, and

$$q(0,x) = 0$$

$$m(0,x) = 0$$

$$n(0,x) = x.$$

By the generalization of Example 11.15 for the functions of three variables, we know that $q(t,x)$, $m(t,x)$, $n(t,x)$ are all p.r.f. of two variables.

Suppose that at moment t_0, M enters the final state 1 for the first time, i.e. $t_0 = \mu_t(q(t,x) = 1)$. Then the output coding of M is $n(t_0,x)$. So the coding of the computed result $M(w)$ for the word $C^{-1}(x)$ is

$$f(x) = C(M(C^{-1}(x))) = n(\mu_t(q(t,x) = 1),x).$$

It should be evident that $f(x)$ is recursive. Since the function $f(x)$ is just the function computed by the Turing machine, so we have

<u>THEOREM 12.1</u> Any Turing-computable function is recursive.

§13 COMPUTING RECURSIVE FUNCTIONS BY TURING MACHINES

In this section, it will be proved that for each recursive function $f(x_1,x_2,\ldots,x_n)$, there is some Turing machine computing it.

Similarly, between a natural number and a word over some alphabet there is an encoding problem. Here, the unary coding is adopted, that is, a Turing machine is constructed having n input tapes and one output tape such that for each $x_1,x_2,\ldots,x_n \in N$,

$$M(a^{x_1}, \ldots, a^{x_n}) = \begin{cases} a^y & \text{if } y = f(x_1, \ldots, x_n); \\ \text{undefined} & \text{if } f(x_1, \ldots, x_n) \text{ is not defined.} \end{cases}$$

(1) Evidently, in unary, there are Turing machines computing the Zero function, the Successor function, and the Projection function, respectively.

(2) If $\psi(x_1, \ldots, x_n)$, $X_1(x_1, \ldots, x_m), \ldots, X_n(x_1, \ldots, x_m)$ can be computed by Turing machines in unary, then the function

$$\phi(y_1, \ldots, y_m) = \psi(X_1(y_1, \ldots, y_m), \ldots, X_n(y_1, \ldots, y_m))$$

can be computed by the following Turing machine in unary.

FUNCTION $\phi(y_1, \ldots, y_m)$

BEGIN

   FOR i := 1 TO n DO $x_i := X_i(y_1, \ldots, y_m)$; $\phi := \psi(x_1, \ldots, x_n)$;

   RETURN ϕ

END

(3) If $\psi(x_2, \ldots, x_n)$, $X(y, x_1, \ldots, x_n)$ can both be computed by Turing machines in unary, and $\phi(x_1, \ldots, x_n)$ is primitive recursive in ψ and X, i.e.

$$\phi(0, x_2, \ldots, x_n) = \psi(x_2, \ldots, x_n)$$

$$\phi(x_1+1, x_2, \ldots, x_n) = X(\phi(x_1, \ldots, x_n), x_1, x_2, \ldots, x_n),$$

then the function $\phi(x_1, \ldots, x_n)$ can be computed by the following Turing machine in unary:

FUNCTION $\phi(x_1, \ldots, x_n)$

BEGIN

   z := 0;

   WHILE $z \leq x_1$ DO

     (IF z = 0 THEN $\phi := \psi(x_2, \ldots, x_n)$ ELSE $\phi := X(\phi, z-1, x_2, \ldots, x_n)$; z := z+1)

   RETURN ϕ

END

(4) If $\psi(t,x_1,\ldots,x_n)$ is a total function, and can be computed by a Turing machine in unary, then the function

$$\phi(x_1,\ldots,x_n) = \mu_t(\psi(t,x_1,\ldots,x_n) = 0)$$

can also be computed by the following Turing machine:

FUNCTION $\phi(x_1,\ldots,x_n)$

BEGIN

    t := 0

    WHILE $\psi(t,x_1,\ldots,x_n) \neq 0$ DO t := t+1;

    RETURN t

END.

Thus, we have

<u>THEOREM 13.1</u> For each recursive function, there exists a Turing machine computing it.

<u>THEOREM 13.2</u> A function $f(x_1,\ldots,x_n)$ on N is recursive if and only if there exists a Turing machine computing it in unary. That is, $f(x_1,\ldots,x_m)$ is a r.f. iff there exists a Turing machine M such that $M(a^{x_1},\ldots,a^{x_n}) = a^{f(x_1,\ldots,x_n)}$ (Here, "=" includes the meaning that once either end is defined then the other end is also defined.)

The unary coding seems too harsh. In other coding systems, what relationships exist between Turing computable functions and recursive functions? In fact, as long as the coding is "reasonable", the functions computed by Turing machines are recursive, and vice versa. But what is a "reasonable" coding system?

<u>DEFINITION 13.1</u> Let I be an alphabet, and C a bijective mapping from I^* to N. If there exists a Turing machine M with I as the input alphabet, and

$$M(w) = a^{C(w)} \text{ (for each } w \text{ in } I^*),$$

then the mapping C is a Turing realizable coding or a reasonable coding.

Thus, a Turing realizable coding C over I^* means that C is a bijective mapping from I^* to N, and there exists some Turing machine coverting w to the representation of C(w) in unary for all w in I^*. Obviously the lexicographically ordered coding on I is Turing realizable. It is defined for $I = \{a_1,\ldots,a_k\}$ as

$C(\Lambda) = 0$

$C(wa_i) = k \cdot C(w) + i$ for each w in I^*.

LEMMA 13.1 Suppose C is a Turing realizable coding. Then there exists some "decoding" Turing machine M' with the input alphabet $I = \{a\}$ such that

$M'(a^x) = C^{-1}(x)$ for all x in N.

Proof: let M be a TM realizing the coding. The decoding Turing machine M' is constructed as follows: first, produce a word w in I^*. Having produced the word, simulate M on w, obtaining M(w). Then compare M(w) with the input, if they are the same, halt with the output w; if not, produce the next word, and repeat the cycle.

Therefore any Turing realizable coding system and unary coding system can be transformed into each other by Turing machines which halt for any input. As far as computability is concerned, there is no difference between various reasonable coding systems. Together with Theorem 13.2, we have

THEOREM 13.3 The function $f(x_1,\ldots,x_n)$ on N is recursive if and only if there exists some TM computing it under a Turing realizable coding system.

In the following, we no longer distinguish between the word on a given alphabet and the corresponding number. For instance, a Turing machine with two input tapes and one output tape can be written as a function $M(x,y)$ ($x,y \in I^*$). In this sense Theorem 13.3 can be written as follows:
(1) For each Turing machine M, $M(x_1,\ldots,x_n)$ is recursive.
(2) Each recursive function is of the form of (1).

Using this language the result of the universal Turing machine can be stated as: There exists a binary recursive function $g(x,y)$ such that for each unary recursive function $f(x)$, there is a natural number e satisfying

$$g(x,e) = f(x).$$

We can regard e as a coding of the function f. In the same way we have

THEOREM 13.4 For an arbitrary integer n > 0 there exists an n+1-ary recursive function $g(x_1,\ldots,x_n,y)$ such that for each n-ary recursive function $f(x_1,\ldots,x_n)$ there is a number $e \in N$ satisfying

$$f(x_1,\ldots,x_n) = g(x_1,\ldots,x_n,e).$$

Such a function g is called an (n-ary) universal recursive function.

§14   RECURSIVE AND RECURSIVELY ENUMERABLE LANUAGES

DEFINITION 14.1 Let L be a language over a given alphabet I. L is recursive or decidable, if there is some Turing machine M such that
 (1) I is the input alphabet of M.
 (2) for any input, M eventually halts.
 (3) L(M) = L.

DEFINITION 14.2 Let L be a language over a given alphabet I. L is recursively enumerable, if there is a Turing machine M such that
 (1) M begins with the empty tape, and the output is

   $w_1 * w_2 * w_3 * \ldots,$

   where the symbol $*$ is not in I, $w_1, w_2, w_3, \ldots \in L$ and for any $w \in L$ there is a natural number i such that $w = w_i$;
 (2) once M prints a $*$, its head will never move back to the left of this $*$.

DEFINITION 14.3 A subset S of the natural number set N is recursive or decidable (recursively enumerable), if the set $\{a^n | n \in S\}$ is a recursive language (recursively enumerable language) over the alphabet I = {a}.

THEOREM 14.1 Suppose L is a language over a given alphabet I. Then L is recursively enumerable iff there is a Turing machine M with the input alphabet I such that L(M) = L.

Proof: NECESSITY. Since L is recursively enumerable, there is a Turing machine M' enumerating all the words in L one by one. We now construct M such that L(M) = L. First, M simulates M' for each input word w, enumerating all the words in L on some other tape. Every time a word is enumerated, the machine compares this word with the input w. If they are the same, then enters some final state, else simulates M' again to enumerate the next word. It is evident that L is the language accepted by M.

SUFFICIENCY. Suppose L = L(M), we proceed to construct a Turing machine M' such that M' can enumerate all the words in L. Let $w_1, w_2, w_3, \ldots$ be all the words in I^* lexicographically ordered. M' generates each (i,j) in the following order:

$$(1,1), (1,2), (2,1), (1,3), (2,2), (3,1), (1,4), \ldots .$$

Having generated (i,j), M' generates w_i on some other tape, then simulates M on w_i for j moves. Whenever M accepts some word w_i, M' enumerates it. For each $w \in L$, M' will eventually find that M accepts w. So M' can certainly enumerate w. Obviously, M' is the desired Turing machine.

COROLLARY 14.1 A recursive language is recursively enumerable.

THEOREM 14.2 Suppose L_1 and L_2 are both recursive, then so are $L_1^c = I^* - L_1$ and $L_1 \cup L_2$. Thus, the class of all recursive languages is closed under all the Boolean operations.

THEOREM 14.3 A language L over some given alphabet I is recursive if and only if L and L^c are both recursively enumerable.

Proof: NECESSITY. Because L is recursive, L^c is also recursive. Thus L and L^c are both recursively enumerable.

SUFFICIENCY. Since L and L^c are both recursively enumerable, so there exist two Turing machines, M_1, M_2, such that $L = L(M_1)$ and $L^c = L(M_2)$. Construct a Turing machine M as follows: for any input w, M simulates M_1 and M_2 on w simultaneously. When either of M_1 and M_2 accepts w, M halts, and only if M_1 accepts w, M accepts w. So $L(M) = L(M_1) = L$ and M halts for any input.

THEOREM 14.4 Suppose L_1, L_2 are recursively enumerable languages, then so are $L_1 \cap L_2$, $L_1 \cup L_2$.

The proof is left as an exercise.

THEOREM 14.5 If a Turing machine M, which halts for all the input words, realizes a transformation from I^* to I_1^*, i.e. T's input alphabet is I, and for any $w \in I^*$, $M(w) \in I_1^*$, and L_1 is a recursive language over I_1, then the language,

$$M^{-1}(L_1) = \{w \in I^* | M(w) \in L_1\},$$

is recursive over I.

Proof: let M_1 be a Turing machine with the input alphabet I_1 such that M_1 halts for each input word and $L(M_1) = L_1$. Obviously, $L(M \cdot M_1) = M^{-1}(L_1)$ and $M \cdot M_1$ halts for each input word.

Similarly, we can prove

THEOREM 14.6 Suppose Turing machine M realizes a transformation from I^* to I_1^* (not necessarily halting for every input), and L_1 is recursively enumerable over I_1. Then

$$M^{-1}(L_1) = \{w \in I^* | M(w) \in L_1\}$$

is a recursively enumerable language over I.

In the following we give some examples and counter-examples of recursive (recursively enumerable) languages.

In constructing the universal Turing machine, the notion was introduced of descriptor d_M for an arbitrary Turing machine M, where $d_M \in \{0,1,*,X\}^*$. For convenience, the Turing machines mentioned below are all single-tape Turing machines with the input alphabets containing $\{0,1,*,X\}$.

EXAMPLE 14.1 The set $S = \{n \in N | f_n(x) = g(x,n)$ is a total function$\}$ is not recursively enumerable.

Proof: suppose S is recursively enumerable. It is thus not difficult to prove that there exists a total recursive function $h(y)$ such that $S = h(N)$, i.e. S is the range of h.

Let $\phi(x,y) = g(x,h(y))$, then $\phi(x,y)$ is a binary total recursive function and for any given total recursive function $\psi(x)$ there exists $e \in N$ such that

$$\psi(x) = \phi(x,e).$$

Set $\psi(x) = \phi(x,x) + 1$. Obviously, $\psi(x)$ is a total recursive function. So there exists $e \in N$, such that $\psi(x) = \phi(x,e)$. Substituting e for x, we have

$$\phi(e,e) + 1 = \phi(e,e),$$

a contradiction. So S is not recursively enumerable.

Thus, Example 14.1 can be stated as that the set of all the codings of total recursive functions is not recursively enumerable, or that the set of all total recursive functions is not recursively enumerable.

EXAMPLE 14.2 The language $L_1 = \{d_M | M \text{ accepts } d_M\}$ is not recursive.

Proof: let L be the set of all w that is a coding of a Turing machine. Suppose L_1 is recursive. Then so is the language

$$L_1' = L - L_1 = \{d_M | M \text{ does not accept } d_M\}.$$

Thus, there is some Turing machine M_0 such that $L_1' = L(M_0)$.

There are two cases.

(1) $d_{M_0} \in L_1'$. By the definition of L_1', M_0 does not accept d_{M_0}, so $d_{M_0} \notin L(M_0) = L_1'$, a contradiction.

(2) $d_{M_0} \notin L_1'$. Since $d_{M_0} \in L$, so $d_{M_0} \in L_1$. By the definition of L_1, M_0 accepts d_{M_0}, so, $d_{M_0} \in L(M_0) = L_1'$, a contradiction. Therefore L_1 is not recursive.

Therefore, we can say that the set $L = \{x \in N | g(x,x) \text{ is defined}\}$ is recursively enumerable but not recursive.

Part two Deterministic similarity
4 Complexity of the Turing machine

§15 COMPUTATIONAL COMPLEXITY

We have introduced theoretical computability in Chapter 2 and Chapter 3. The most significant basis in theoretical computability is the Turing-Church thesis, which gives to the intuitive notion of what is "computable" a formal description: according to the Turing-Church thesis, we can reach conclusions such as "this class of problems is computable".

In addition to theoretical computability, we have to consider computational complexity. A problem may be theoretically computable, but its computation may be too complicated (e.g. it will consume too much time) to execute practically.

How can we measure the complexity of a computation?

For a given class of problems, there are infinitely many problems. Usually we associate a natural number n with each problem in order to measure its magnitude. We call this the "size" of the problem. For example, the multiplication of two n-bit numbers is of size n. If we use the Turing machine as the computational model, then every problem to be solved is encoded into a string which is the input to the Turing machine. We define the size of the problem as the length of the input string.

Every computational device has several resources, such as time, space, reversal. In solving problems, i.e. in computing, we consume these resources. For example, computation must consume time. Intuitively, computational complexity is measured by the resources consumed in the computation process.

If we can, within $t(n)$ time units, solve every problem of size n in a given class of problems, then we say that the time complexity of this class of problems is $t(n)$. In the case of the Turing machine, the concept of time complexity can be defined precisely; that is, if a Turing machine halts within $t(n)$ moves, then its time complexity is defined to be $t(n)$ (see next section).

Suppose that a computer program is of time complexity $t(n) = \exp(n)$ and that this computer can complete one million moves per second. Then we have the following table:

n	time
20	1 second
30	17 minutes
40	12.7 days
50	35.7 years
60	36558 years
70	378 million years

Therefore the program works only for very small n. We can say that it is impossible to solve problems by this computer program in practice.

Usually, problems of polynomial complexity are considered to be computable in practice, while those of exponential complexity are not computable in practice. Similarly, if the complexity functions of two Turing machines are polynomially related (see below for the precise definition), then their complexities are considered to be essentially the same.

For convenience, we give the following definitions.

DEFINITION 15.1 Let $f(n)$, $g(n)$ be positive value functions in the set N of natural numbers.
  (1) If there is a constant C such that $f(n) \leq C \cdot g(n)$ for all n, then we write
    $$f(n) = O(g(n))$$
  (2) If $f(n)/g(n) \to 0$ $(n \to \infty)$, then we write
    $$f(n) = o(g(n))$$
  (3) If $g(n) = O(f(n))$, then we write
    $$f(n) = \Omega(g(n))$$
  (4) If $f(n) = O(g(n))$ and $g(n) = O(f(n))$, then we write
    $$f(n) = \Theta(g(n))$$
  (5) If there is a polynomial p such that $g(n) = p(f(n))$, then we write
    $$g(n) = f^*(n)$$

(6) If there is a polynomial p such that $g(n) \leq p(f(n))$, then we write
$$g(n) \leq f^*(n)$$
(7) If $f(n) \leq g^*(n)$ and $g(n) \leq f^*(n)$, then we write
$$g(n) \sim^* f(n)$$

EXAMPLE 15.2 $n \sim^* n^3$, $n^2 \sim^* n^4$, $2^n \sim^* 3^n$.

Exercises

15.1 For positive value functions prove that
  (1) If $f_1 = O(g_1)$, $f_2 = O(g_2)$, then $f_1 + f_2 = O(g_1 + g_2)$, $f_1 \cdot f_2 = O(g_1 \cdot g_2)$
  (2) If $f_1 = \Omega(g_1)$, $f_2 = \Omega(g_2)$, then $f_1 + f_2 = \Omega(g_1 + g_2)$, $f_1 \cdot f_2 = \Omega(g_1 \cdot g_2)$,
  $f_1 + f_2 = \Omega(\max(g_1, g_2))$

15.2 For constant $c > 1$ and $k > 0$ prove that
  (1) $c^n \sim^* 2^n$
  (2) $n \sim^* n^t$

15.3 Prove that $(n \sim^* 2^n)$ does not hold.

15.4 Find a positive value function f such that
  (1) $f(n)$ is an increasing function
  (2) For any positive constant k, $f(n) \neq O(n^k)$
  (3) For any positive constant k, $f(n) \neq \Omega(n^k)$

§16 RESOURCES OF THE MULTITAPE TURING MACHINE

From now on, unless we state otherwise, all Turing machines satisfy:

(1) There is one read-only input tape.

(2) The input tape head never goes beyond the input word, that is, it always scans a symbol of the input word or the two blank symbols adjacent to the input word.

(3) There are k work tapes, where $k > 1$.

(4) There is one write-only output tape and the output tape head never moves to the left.

Thus the contents of the input tape remain unchanged during the whole computation. It is installed to provide the input information, i.e., the problem that is to be computed. So it is of no use to move the input tape head far away from the input word. The output tape records the ultimate result of the computation, so it is write-only and its head can only move to the right or remain stationary. The work tapes are the storage of the Turing machine. They store the intermediate results of the computation for further use. So they are read-write tapes and their heads can move to the left or the right.

DEFINITION 16.1 Let M be a Turing machine and w be an input word. We use $t(w)$ to denote the total number of steps of M for input w ($t(w)$ may be ∞) and say that the time consumption of M for input w is $t(w)$. Function

$$t(n) = \max \{t(w) \mid |w| \leq n\}$$

is called the time complexity of M.

Obviously, $t(n)$ is a non-decreasing function. $t(n) = n^2$ means that:

(1) M halts for all inputs.

(2) For all inputs of length less than or equal to n, M halts within $t(n)$ steps.

(3) There exists an input word w of length less than or equal to n such that M does exact n^2 moves.

$t(n) = O(n^2)$ means that there exists a constant $c > 0$ such that M halts within $c \cdot n^2$ steps for all inputs of length less than or equal to n.

DEFINITION 16.2 Let M be a Turing machine and w be an input word. We use $s(w)$ to denote the number of the work tape squares that have been scanned during the whole computation of M for the input word w ($s(w)$ may be ∞) and say that the space consumption of M for input w is $s(w)$. The function

$$s(n) = \max \{s(w) \mid |w| \leq n\}$$

is called the space complexity of M.

Similarly s(n) is a non-decreasing function.

Notice that s(w) is the number of squares on work tapes only. The squares on input tape and output tape that have been scanned are not included. So s(w) may be less than $|w|$ or $|M(w)|$. Therefore it is possible, for some Turing machine M, that $s(n) = O(\log n)$, while $|M(w)| = n^2$ for some input w of length n.

In order to introduce two important but slightly complicated notions, phase and reversal, we have to discuss the procedure of computation in more detail.

Let M be a Turing machine; the computation of M for an input word w may be described by an ID sequence $ID_0 \vdash ID_1 \vdash ID_2 \vdash \ldots$, where ID_0 is the initial ID of M for w. In one step $ID \vdash ID'$, every tape head will move at most one square in direction d (d = R, L or S). We say that R and L are different directions. If, in two successive steps $ID \vdash ID' \vdash ID''$, a work tape head moves in different directions, then we say that this head changes its movement direction during the period $ID \vdash ID' \vdash ID''$.

<u>DEFINITION 16.3</u>   Let $ID_0 \vdash ID_1 \vdash ID_2 \vdash \ldots$ be the computation of a Turing machine M for an input word w. We say that interval (i,j) is a phase of this computation (i < j, j may be ∞) if no work tape head changes its movement direction during

$$ID_i \vdash ID_{i+1} \vdash ID_{i+2} \vdash \ldots \vdash ID_j.$$

If $(0,\infty)$ can be divided into r disjoint intervals, each of which is a phase of the above computation, then we say that r is the reversal number of this partition (r may be ∞). The minimal reversal number of M for input w is denoted as r(w). The function

$$r(n) = \max \{r(w) \mid |w| \leq n\}$$

is called the reversal complexity of M.

Notice that the input tape head may change its movement direction in one phase, i.e., the change of input head direction does not interrupt a phase. This is reasonable because the change of the direction of the work tape heads means that intermediate results are used, but the change of the direction of the input tape head does not. From this point of view, we may say that the reversal complexity of M tells us how many times M makes use of the intermediate results.

We have introduced three main resources for the multitape Turing machine - time, space, reversal - and their corresponding complexities. Throughout this chapter, we always use $t(n)$, $s(n)$ and $r(n)$ to denote respectively the time complexity, space complexity and reversal complexity of a Turing machine. If we discuss several Turing machines M_i ($i \in I$) at the same time, then the functions $t_i(n)$, $s_i(n)$, $r_i(n)$ are used to represent the complexity functions of M_i ($i \in I$).

From now on, we only discuss Turing machines with time complexity $t(n) < \infty$, i.e., the languages accepted by them are recursive. Hence we have $s(n) < \infty$ and $r(n) < \infty$.

All the above complexities are defined for Turing machines. For a given class of problems (e.g., recognizing a language L), one may design many different algorithms, i.e., Turing machines. These Turing machines, generally speaking, have different complexity functions. How can we measure the complexity of this class of problems (i.e., language)?

If a Turing machine M solves a class of problems and, say, the time complexity of M does not exceed $f(n)$, then we say that $f(n)$ is the upper bound of the time complexity of this class. If the time complexity of every Turing machine solving this class of problems is at least $f(n)$, then we say that $f(n)$ is the lower bound of the time complexity of this class.

Similarly we can define the space (reversal) upper and lower bounds of a class of problems.

EXAMPLE 16.1 (converting a unary number to a binary number).

INPUT: a^n, where n is a natural number.

OUTPUT: $b_0 b_1 b_2 \ldots b_m$ where $b_i \in \{0,1\}$, $\sum_{o}^{m} b_k \cdot 2^k = n$ and $b_m = 1$ if $m \neq 0$.

(1) Design a Turing machine with $r(n) = O(\log n)$ and $s(n) = O(n)$.

Notice that $b_0 = MOD(n,2)$, $b_1 = MOD([n/2],2),\ldots,$where $MOD(n,m)$ is the remainder of n divided by m. We design a Turing machine M with two work tapes X and Y. M does the following work.

1. If the input tape is empty, i.e., the input tape head reads a blank

symbol ␣ at moment t = 0, then M writes a symbol 0 on its output tape and halts.

2. M copies the contents of the input tape, that is, a^n, to its first work tape X. Now the head of tape X is on the rightmost position of the contents of X.

3. M moves the head of tape X from the right to the left and deletes a's from X. For two a's deleted from X, it writes one symbol a on Y until there are less than two a's on tape X. Then M writes a symbol 1 or 0 on its output tape according to whether the number of a's on X is odd.

4. If Y is empty then M halts. If not, then M copies the contents of Y to X, clears Y and GOTO step 3.

Obviously step 3 writes [m/2] many a's on Y if there are m a's on X and does not interrupt a phase. This step will be repeated log n times, so

$r(n) = O(\log n)$.

Since no work tape uses more than n squares, we have

$s(n) = O(n)$.

This Turing machine can be described as

```
BEGIN
  IF  input=␣THEN write 0 ELSE        [input=␣ means n = 0]
    BEGIN
     X:=input;                        [copy input tape to tape X]
     REPEAT
      Y:=[X/2];
      write MOD(X,2);
      X:=Y
     UNTIL Y=0
    END
END.
```

(2) Design a Turing machine with $s(n) = O(\log n)$ and $r(n) = O(n)$.

If there is a binary number on a work tape of a Turing machine M, then M can compute the successor (plus 1) in one phase without using other tapes. The desired Turing machine may be designed as follows.

1. Write 0 on its work tape.
2. Moving its input tape head from the left to the right, the TM counts the number of symbols on its input tape.

After M finishes the input reading, the binary representation of n will be obtained on its work tape. Because the length of binary number n is $O(\log n)$, we have

$$s(n) = O(\log n).$$

The "Plus 1" operation will be done n times, so we have

$$r(n) = O(n).$$

EXAMPLE 16.2 (generate all unary numbers not exceeding n).

INPUT: a^n, where n is a natural number
OUTPUT: $*a*a^2*a^3* \ldots *a^n*$
COMPLEXITY: $r(n) = O(1)$, $s(n) = O(n^2)$, $t(n) = O(n^2)$

First M writes the following contents on its work tapes in $O(1)$ phases, using the input tape head as a counter.

first work tape: $(a^{n+1}\#)^n a^{n+1}$ = a...a#a...a#.........#a...a
second work tape: $(*a^{n+2})^n *$ =*a....a*a....a*.....*a....a*

Notice that the strings on the two work tapes are of the same length.

Then both work tape heads move to the leftmost ends and begin scanning from the left to the right synchronously. Meanwhile, M copies the letters on the second work tape to the output tape until a letter * has been written, and repeats the above copy procedure whenever the first work tape head passes a letter #.

As the last example of this section, we design a Turing machine to determine whether there is a cycle in an undirected graph.

For an (undirected) simple graph G, we denote the vertices of G as $1,\ldots,|V|$, each of which is in binary form. The adjacency list of vertex v is denoted as

$$(v,v_1,v_2,\ldots,v_k)$$

where $v_1, v_2, \ldots, v_k$ are all the vertices adjacent to v. The concatenation of the adjacency lists of all the vertices of G is called the coding of G.

EXAMPLE 16.3 (cycle detecting problem for an undirected graph).

INPUT: the coding of a simple undirected graph G = (V,E)

OUTPUT:1 (when there is a cycle in G) or 0 (otherwise)

We give an s(n) = O(log n) algorithm (Turing machine) which was first presented, as the "Three Chinese Algorithm", by the author in 1980 for the study of the complete problem of deterministic space. Here we give a slightly revised version which may be called the "Two Chinese Algorithm", because only two Chinese are involved.

Obviously, if n is the length of the input, then $|V| \leq n$ and $|E| \leq n$.

Imagine that two Chinese - father and son - are travelling along the edges of G. Father sits at the vertex v (v = 1,2,...,$|V|$) while the son walks by the so-called "cyclic searching principle":

* for every vertex u of G, give all the edges out from u an order according to the magnitudes of their codings: thus we can say "the first edge of u", "the second edge of u", and so on. Edge (u,v) may be the i-th edge of u and the j-th edge of v with i ≠ j.

* whenever entering a vertex u of fanout k along edge i (of u), one must go out from u along edge i + 1 (mod k), (edge k + 1 is edge 1).

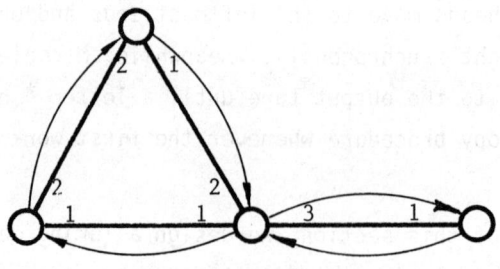

Fig. 16.1 Cyclic searching

Father remembers the edge along which the son departed and sees if he comes back along the same edge. If, for every edge adjacent to v, the son does so, then the father will take his son to vertex v+1 (if v<$|V|$) or claim that G has no cycle (if v = |V|). Otherwise there must be a cycle.

It is not difficult to prove that:

(1) G has no cycle iff, for every vertex, the son always returns by the same edge along which he departed.

(2) starting from vertex v, the son will surely come back to v before he passes $2 \cdot |E|$ edges.

The correctness of this algorithm is guaranteed by the above two statements.

All the intermediate information can be stored in $O(\log n)$ space. So

$s(n) = O(\log n)$.

In the rest of this section, we will prove two important theorems - linear speed-up theorems - which tell us that the constant factor of the time complexity or space complexity is not substantial.

THEOREM 16.1 Let M be a Turing machine with space complexity $s(n) \to \infty (n \to \infty)$, then, for any $\varepsilon > 0$, there exists another Turing machine $\hat{M}_1$ simulating M and satisfying

$s_1(n) \leq \varepsilon \cdot s(n)$ (for sufficiently large n)

where simulation means that the machines give the same output for the same input.

Proof: construct $\hat{M}_1$ as follows.

Choose a positive integer m such that $1/m < \varepsilon$.

(1) The input tape and output tape of M_1 are just the same as those of $\hat{M}$.

(2) Each square of the k work tapes of M_1 stores the contents of m consecutive squares on the corresponding tape of $\hat{M}$, i.e., every work symbol of M_1 consists of m consecutive symbols of M.

(3) M_1 uses its state to "remember" the state of M and the position of each work tape head of M in the m consecutive squares which is one symbol on the work tapes of M_1.

Thus Turing machine M_1 knows the state of M and the symbols scanned by the heads of M at any moment. According to the next move function of M, M_1 changes its own state, rewrites the contents of its work tape squares, writes on output tape and moves its tape heads.

The detailed construction of M_1 is straightforward but tedious. We leave it to the reader.

Obviously $s_1(n) \leq [s(n)/m] + 2K$. Since $s(n) \to \infty$, we have $s_1(n) \leq \varepsilon \cdot s(n)$ for sufficiently large n.

According to Theorem 16.1, space complexity can be compressed by a constant factor. For example, if a language L can be recognized by a Turing machine of space complexity $100 \cdot s(n)$, then it can be recognized by another Turing machine of space complexity $s(n)$. That is, it is the growth rate of complexity function that is important. We shall say the space complexity of a Turing machine is, say, log n instead of O(log n).

THEOREM 16.2 Let M be a Turing machine with time complexity $t(n) \to \infty$ $(n \to \infty)$, then, for any $\varepsilon > 0$, there exists another Turing machine M_1 simulating M and satisfying

$$t_1(n) \leq n + \varepsilon \cdot (t(n) + n) \quad \text{(for sufficiently large n).}$$

Proof: the input symbols of M and M_1 are the same. Every work symbol of M_1 is m consecutive symbols of M. Thus every square of work tapes of M_1 is able to store the contents of m consecutive squares (including input tape and output tape) of M. If M has k work tapes, then M_1 will use k+1 work tapes. The first work tape is used to simulate the input tape of M.

M_1 can remember the following information by its finite control:

(1) the current state of M;

(2) k+1 triples of "big" symbols of M_1. Every triple consists of 3 consecutive work tape symbols of M_1, i.e., $3 \cdot m$ consecutive symbols of the input tape or one of k work tapes of M with the heads of M scanning one "small symbol" in the middle "big symbol";

(3) the positions of the heads of M excluding the output tape head; and

(4) i output symbols of M that have not been written out by M_1, where i < m.

M_1 simulates M as follows:

STEP 1. M_1 copies the input w into its first work tape, encoding m consecutive symbols into one work symbol.

STEP 2. M_1 moves its first work tape head to the leftmost symbol.

STEP 3. After reading the squares adjacent to the scanning ones on the work tapes by four moves, M_1 has all the information needed for simulating m moves of M. Then M_1 makes two moves to simulate m moves of M. Whenever the number of output symbols remembered by the state of M_1 is greater than or equal to m, then M_1 writes one 'big' output symbol on its output tape, leaving the remainders in its finite control.

The detailed design of the next move function of M_1 is left to the reader.

Step 1 will consume time n+1. Step 2 will consume time $\lceil n/m \rceil + 1$. In Step 3, every six moves of M_1 can simulate m moves of M. Hence $t_1(n) \leq \lceil 6 \cdot t(n)/m \rceil + \lceil n/m \rceil + 2+n$. If we choose m so that $6/m < \varepsilon$, then $t_1(n) \leq n + \varepsilon \cdot (t(n) + n)$ for sufficiently large n.

COROLLARY 16.1 Let M be a Turing machine with time complexity t(n) such that $t(n)/n \to \infty$; then, for any $\varepsilon > 0$, there exists another Turing machine M_1 simulating M and

$$t_1(n) \leq \varepsilon \cdot t(n) \quad \text{(for sufficiently large n)}.$$

Notice that the outputs of M_1 and M are different from each other. Suppose that the output of M is of length ℓ. Then the output of M_1 is of length $\lceil \ell/m \rceil$. If we introduce the restriction that the simulator must have the same output as M does, then, if the output of M is of length t(n), the time complexity of every simulator is at least t(n) and the time compression becomes impossible.

Exercises

16.1 Show that if a Turing machine M is of space complexity O(1), then L(M) is regular.

16.2 Show that if r(n) = 1 for a Turing maching M, then L(M) is regular.

16.3 Design a Turing machine M such that r(n) = 2 and L(M) is not regular.

16.4 Prove that t(n), s(n) and r(n) are all recursive functions.

16.5 Show that for any Turing machine M, there exists another Turing machine M_1 such that
(1) M and M_1 have exactly the same outputs for the same inputs;

(2) M and M_1 have exactly the same time, space and reversal complexity and

(3) M_1 does not move its work tape or output tape head until it rewrites some new symbol in the currently scanned square.

16.6 Let M be a Turing machine and w be an input. Choose $x_0 = 0$. For $i = 0,1,2,\ldots$, suppose that an integer x_i has been chosen. If $x_i < t(w)$, choose x_{i+1} as large as possible under the condition that the interval (x_i, x_{i+1}) is still a phase. Suppose that $x_r = t(w)$. Prove that $r = r(w)$.

Design Turing machines satisfying the conditions in Exercises 16.7 - 16.16.

16.7 INPUT: $a_1 a_2 \ldots a_n$, where $a_i \in \{0,1\}$

OUTPUT: $\#a_1 \# a_1 a_2 \# a_1 a_2 a_3 \# \ldots \ldots \# a_1 a_2 \ldots a_n \#$

(1) COMPLEXITY: $r(n) = O(1)$
$s(n) = O(n^2)$

(2) COMPLEXITY: $r(n) = O(n^2)$
$s(n) = O(\log n)$

16.8 INPUT: a^n

OUTPUT: all the strings of length n on $\{0,1\}$ with a symbol # inserted between any two adjacent strings

(1) COMPLEXITY: $r(n) = O(n)$
$s(n) = O(n \cdot 2^n)$

(2) COMPLEXITY: $r(n) = O(2^n)$
$s(n) = O(n)$

16.9 INPUT: binary natural number x

OUTPUT: binary natural number $\lceil \log(x+1) \rceil$

(1) COMPLEXITY: $r(n) = O(\log n)$
$s(n) = O(n)$
$t(n) = O(n \cdot \log n)$

(2) COMPLEXITY: $r(n) = O(n)$
$s(n) = O(\log n)$
$t(n) = O(n \cdot \log n)$

16.10 INPUT: unary natural number x

OUTPUT: unary natural number $\lceil \log(x \div 1) \rceil$
- (1) COMPLEXITY: $r(n) = O(\log n)$
  $s(n) = O(n)$
  $t(n) = O(n \cdot \log n)$
- (2) COMPLEXITY: $r(n) = O(n)$
  $s(n) = O(\log n)$
  $t(n) = O(n \cdot \log n)$

16.11 INPUT: $x \# a^k$, where $x \in \{0,1\}^*$

OUTPUT: x^k

COMPLEXITY: $r(n) = O(1)$
$s(n) = O(n)$
$t(n) = O(n^2)$

16.12 INPUT: $x \# y$, where x and y are binary natural numbers

OUTPUT: binary natural number $x + y$

COMPLEXITY: $r(n) = O(1)$
$s(n) = O(n)$
$t(n) = O(n)$

16.13 INPUT: $x_1 \# x_2 \# \ldots \# x_k$, where x_i's are binary natural numbers

OUTPUT: binary natural number $x_1 + x_2 + \ldots + x_k$

COMPLEXITY: $r(n) = O(\log n)$
$s(n) = O(n)$
$t(n) = O(n \cdot \log n)$

16.14 INPUT: $x \# y$, where x and y are binary natural numbers

OUTPUT: binary natural number $x \cdot y$
- (1) COMPLEXITY: $r(n) = O(\log n)$
  $s(n) = O(n^2)$
  $t(n) = O(n^2 \cdot \log n)$
- (2)* COMPLEXITY: $s(n) = O(\log n)$

16.15 INPUT: $x_1 \# x_2 \# \ldots \# x_k$, where x_i's are binary natural numbers

OUTPUT: binary natural number $\text{MAX}\{x_i ; 1 \leq i \leq k\}$
- (1) COMPLEXITY: $r(n) = O(n)$
  $s(n) = O(n)$
- (2) COMPLEXITY: $s(n) = O(\log n)$

16.16 INPUT: $x_1 \# x_2 \# \ldots \# x_k$, where x_i's are binary natural numbers
OUTPUT: $y_1 \# y_2 \# \ldots \# y_k$, where y_i's are x_i's and $y_1 \leq y_2 \leq \ldots \leq y_k$
COMPLEXITY: $s(n) = O(\log n)$

16.17 Prove the correctness of the algorithm in Example 16.3.

16.18 A one-directional TM is a multitape TM satisfying that
(1) in odd phases all its input tape and work tape heads move from left to right;
(2) in even phases all its input tape and work tape heads move from right to left, write nothing on the work tapes, and come back to the left-hand ends of the corresponding tapes.

A multitape TM is oblivious if the positions of its tape heads at time t can be completely determined by t and the input length n (having nothing to do with the input contents)

Prove that there is an oblivious one-directional TM to duplicate the input word w to n (= |w|) copies w^n within $O(\log n)$ reversal and $O(n^2)$ space.

16.19 Design an oblivious one-directional TM to reverse the input word w within $O(\log n)$ reversal and $O(n)$ space.

16.20 Suppose that there is an interval of length n on a work tape. In this interval there is a word w. Design an oblivious one-directional TM moving the word w to the left-hand end of the interval within $O(\log n)$ reversal and $O(n)$ space.

16.21* Given a word $w \in \{0,1,\emptyset\}$, if we remove all occurrences of $\emptyset$, then we obtain a word $\bar{w} \in \{0,1\}$, called the projection of w. Design an oblivious one-directional TM to compute the projection of the input $w \in \{0,1,\emptyset\}$ within $O(\log^2 n)$ reversal and $O(n)$ space.

16.22* For any multitape TM T of reversal $r(n)$ and space $s(n)$, there is a one-directional TM simulating T within $O(r(n)(\log s(n) + \log n))$ reversal and $O(s(n) + n)$ space. If T is oblivious, then so is the simulator. [Hint: use Exercise 16.19.)

16.23* (Pippenger) Suppose that T is a TM of space complexity $s(n)$. Then there is an oblivious TM simulating T within $O((s(n) + \log n)^2)$ reversal and $O(n^3 c^{s(n)})$ space.

16.24* A TM with f reversal and g space can be simulated by an oblivious one-directional TM with f* reversal and g* space.

§17 BASIC RELATIONSHIPS AND PROPERTIES

Throughout this section, M is a Turing machine with k work tapes. Sometimes, instead of $r(n)$, $s(n)$ and $t(n)$, we use r, s and t to denote respectively the reversal, space and time complexity of M.

Obviously we have

$$s(n) = O(t(n))$$
$$r(n) = O(t(n)).$$

Since it is assumed in this book that $t(n) < \infty$, we have $s(n) < \infty$ and $r(n) < \infty$.

For a given input w of length n, the state of the input tape at any moment can be completely described by the position of the input tape head, because of the 'read-only' attribute. This position is a natural number j indicating that the input tape head is scanning the j-th symbol of w, where $0 \leq j \leq n+1$ (the 0-th and the (n+1)-th symbols are both ⊔). The state of the i-th work tape at any moment can be completely described by $x_i \Delta y_i$, where $\Delta \notin \Sigma$, indicating that the contents of this tape are $x_i y_i$ and the head is scanning the first symbol of y_i. If the state of the FC at this moment is $q \in Q$, then

$$D = (q, j, x_1 \Delta y_1, x_2 \Delta y_2, \ldots, x_k \Delta y_k) \text{ where } q \in Q \qquad (17.1)$$

differs from ID only in that the contents of input tape and output tape are not specified. We call (17.1) the configuration of M at this moment. Obviously the computation of M for w may be considered as a chain of configurations:

$$D_0 \vdash D_1 \vdash \ldots D_t.$$

Because $|x_i| + |y_i| \leq s = s(n)$, there are at most

$$|Q| \cdot (n+2) \cdot (s(n) \cdot |\Sigma|^{s(n)})^k$$

many configurations for input w. Since $t(n) < \infty$, these D_i's are different from each other. So we have

THEOREM 17.1 There exists a constant $c > 1$ such that

$$t(n) = O(n \cdot c^{s(n)}).$$

Notice that within a phase the movement direction of every work tape head remains unchanged. So, for a given moment, if we know the state q of M, the position j of the input tape head and the k work tape contents $z_1, z_2, \ldots, z_k$ ahead of their heads (in the movement direction), the movements of M from this moment to the end of the current phase are uniquely determined. We call

$$(q, j, z_1, z_2, \ldots, z_k) \qquad (17.2)$$

the phase-configuration of M at this moment.

Notice that the length of z_i decreases within the current phase. If all the work tape heads remain stationary, there are at most $|Q| \cdot (n+2)$ many different phase-configurations. So the sum of $|z_i|$ decreases by at least 1 within every $|Q| \cdot (n+2)$ steps. That is, starting from the phase-configuration (17.2), M will surely enter the next phase within $O(n \cdot (1 + \Sigma |z_i|))$ steps.

Suppose that the phase-configuration (17.2) is in phase $m+1$. Obviously all the symbols of these z_i's are written on the work tapes in previous phases, that is, from phase 1 to phase m. Let the total space consumption in these m phases be $s(w,m)$ and the time consumption in phase m be $t(w,m)$. We have

$$\Sigma |z_i| \leq s(w,m)$$

$$t(w, m+1) = O(n \cdot (1 + \Sigma |z_i|)) = O(n \cdot s(w,m) + n) \qquad (17.3)$$

$$t(w, 1) = O(n).$$

So

$$t(w) = t(w,1) + t(w,2) + \ldots + t(w, r(w))$$

$$= O(n + (n \cdot s(w,1) + n) + \ldots + (n \cdot s(w, r(w)-1) + n))$$

$$= O(n \cdot (s(n) + 1) \cdot r(n)).$$

Thus we have

THEOREM 17.2 If $s(n) \geq 1$ then $t(n) = O(n \cdot s(n) \cdot r(n))$.

Since at most $k \cdot t(w,m+1)$ squares on work tapes are scanned in phase $m+1$, we have

$$s(w,m+1) \leq k \cdot t(w,m+1) + s(w,m)$$

$$= O(n \cdot s(w,m) + n) + s(w,m)$$

$$\leq c \cdot n(s(w,m) + 1) \qquad \text{where c is a positive constant.}$$

Substituting repeatedly, we have

$$s(w,m) \leq cn(s(w,m-1) + 1)$$

$$\leq cn(cn(s(w,m-2) + 1) + 1)$$

$$\leq \ldots \leq cn + (cn)^2 + \ldots + (cn)^m \leq (c_1 n)^m.$$

By (17.3) we have

$$t(w,m) = O(n \cdot s(w,m-1) + n)$$

$$= O(n((c_1 n)^{m-1} + 1))$$

$$= O((c_1 n)^m).$$

Therefore we obtain

$$t(w) = t(w,1) + t(w,2) + \ldots + t(w,r(w))$$

$$= O(c_1 n + (c_1 n)^2 + \ldots + (c_1 n)^{r(w)}) \leq (c_2 n)^{r(w)}$$

where c_1 and c_2 are constants.

THEOREM 17.3 There exists a constant $c > 1$ such that

$$t(n) = O((c \cdot n)^{r(n)}).$$

Thus we have the following relations among $r(n)$, $s(n)$ and $t(n)$:

$$s(n) = O(t(n)) \tag{17.4}$$

$$r(n) = O(t(n)). \tag{17.5}$$

$$t(n) = O(n \cdot s(n) \cdot r(n)) \tag{17.6}$$

$$t(n) = O(n \cdot c^{s(n)}) \tag{17.7}$$

$$t(n) = O((c \cdot n)^{r(n)}) \tag{17.8}$$

Because of (17.6), we will mainly discuss $r(n)$ and $s(n)$.

THEOREM 17.4 Let M_1 and M_2 be two Turing machines such that

$$s_1(n) = g(n)$$

$$s_2(n) = O(\log n)$$

Then there exists a Turing maching M simulating $M_1 \cdot M_2$ (for the definition of $M_1 \cdot M_2$, refer to section 9) such that the space complexity $s(n)$ of M satisfies

$$s(n) = O(g(n) + \log n).$$

Proof: because of (17.7) the length of the output of M_1 can be represented by a binary number of at most $O(g(n) + \log n)$ bits.

The desired Turing machine M can be designed as follows.

M simulates M_2 on some work tapes step by step. Whenever M_2 needs the j-th input symbol, M stops simulating M_2 and starts simulating M_1 on other work tapes from the beginning without writing down the output of M_1 until M gets the j-th output symbol of M_1, i.e., the j-th input symbol of M_2. With this symbol at hand, M resumes simulating M_2 for the next step.

The space for simulating M_1 is $g(n)$. Since the length of the output of M_1, i.e., the length of the input of M_2, is $O(n \cdot c^{g(n)})$, the space for simulating M_2 is $O(\log(n \cdot c^{g(n)})) = O(g(n) + \log n)$.

As well as the space simulating M_1 and M_2, M needs some extra space for some "pointers" to indicate the position of the input tape head of M_2. The space used for these pointers is $O(g(n) + \log n)$. Therefore the total space used is

$$s(n) = O(g(n) + \log n)$$

COROLLARY 17.1   If both M_1 and M_2 have space complexity $O(\log n)$, then there is a Turing machine M simulating $M_1 \cdot M_2$ and having space complexity $O(\log n)$.

There are many tradeoffs in the phenomenon of computation. Rackoff and Dymond noticed one kind of tradeoff phenomenon, space versus reversal. Except in some very simple languages, if one wants to reduce the space, then one has to use more reversals, and vice versa. Rackoff and Dymond conjectured that if $r(n) \cdot s(n) = o(n)$, then the language accepted by the Turing machine is regular. The author gives this conjecture a positive answer in a symmetric form.

THEOREM 17.5   If $r(n) \cdot s(n) = o(n)$, then $r(n) \cdot s(n) = O(1)$.

Proof: since the number of the work tape head movements in any phase is less than or equal to $s(n)$, the total number of movements of work tape heads during the whole computation is less than or equal to $r(n) \cdot s(n) = o(n)$.

Let w be any input of length n. We call the i-th input square "marked" if there is a moment at which the input head is scanning the i-th input square and at least one work tape head is going to move left or right. Let I be an interval in the input tape in which no square is marked. In other words, whenever the input head is inside I, the work tape head does not move at all.

Now consider a pair of squares (i,j) in I, where $i < j$. Suppose that at some stage of the computation the input head is at square i-1, M is in state q and the k work tape heads are scanning symbols $a_1, \ldots, a_k$ respectively. Suppose also that the input head moves one square to the right, and the input head does not leave the interval (i,j) until it comes back to i-1 (leaving the interval from the left-hand side); this time M enters state p, and the work tape heads scan symbols $b_1, \ldots, b_k$ respectively (notice that these work tape heads do not move at all during this period). We use the following notation to express this situation:

$$\text{left}, q, a_1, \ldots, a_k \rightarrow \text{left}, p, b_1, \ldots, b_k$$

If the input head leaves the interval (i,j) from the right-hand side, that is, it comes to j+1, we express this by

$$\text{left, } q, a_1, \ldots, a_k \rightarrow \text{right, } p, b_1, \ldots, b_k$$

In the same way, if the input head enters (i,j) from the right-hand side, and leaves from the left or the right, we write

$$\text{right, } q, a_1, \ldots, a_k \rightarrow \text{left, } p, b_1, \ldots, b_k$$

$$\text{right, } q, a_1, \ldots, a_k \rightarrow \text{right, } p, b_1, \ldots, b_k$$

respectively.

All these notations constitute a characteristic table of (i,j). Obviously, there is only a finite number, say c, of different characteristic tables.

We prove that for any input word w, if there is an interval of length c+1 in which there is no marked square, then we can find a shorter input v which uses the same number of work tape movements as w.

Assume that this interval is (i,i+c). Consider the following c+1 different intervals: (i,i), (i,i+1),...,(i+c). There must be at least two integers $j < \ell$, $i \leq j < \ell \leq i + c$, such that the following two intervals have the same characteristic table:

$(i,j), (i,\ell)$

Suppose that the input is $w = a_1 a_2 \ldots a_n$. Set

$$v = a_1 a_2 \ldots a_i \ldots a_j a_{\ell+1} \ldots a_n$$

It is easy to see that w and v have the same worktape movements. Now set

$$L_m = \{w | M \text{ makes exactly m movements of the work tape heads on } w\}$$

If for sufficient large m, L_m is empty, then we already have $r(n) \cdot s(n) = O(1)$. Therefore, we consider the case that there are infinitely many m such that L_m is not empty. For these m define

$$\ell_m = \min \{|w| \mid w \in L_m\}$$

We will show that $\ell_m \leq (c+1) \cdot (m+1)$.

In fact, by m movements of the work tape heads, the input is divided into at most m+1 intervals containing no marked squares. If $\ell_m > (c+1) \cdot (m+1)$, then there must be an interval of length $\geq c+1$ in which there is no marked square. As above, we can shorten this to obtain a new input which is still in L_m, a contradiction to the definition of ℓ_m.

Remember that $r(n) \cdot s(n)$ is not less than the total number of work tape movements. If w is the shortest member in L_m and $|w| = n$, then we have

$$r(n) \cdot s(n) \geq m \geq (\ell_m/(c+1))-1 = |w|/(c+1)-1 = n/(c+1)-1$$

for infinitely many m. This contradicts the fact that $r(n) \cdot s(n) = o(n)$. So L_m must be empty for sufficiently large m and we have

$$r(n) \cdot s(n) = O(1).$$

Exercises

17.1 Design Turing machines satisfying each of the following conditions.
  (1) $s(n) \to \infty$ and
  $$t(n) \geq n \cdot 2^{s(n)}$$
  (2) $r(n) \to \infty$ and
  $$s(n) \geq n^{r(n)}$$
  (3) $r(n) \to \infty$, $s(n) \to \infty$ and $t(n) \geq n \cdot s(n) \cdot r(n)$

17.2 Prove that there is no such Turing machine M that
  (1) M calculates the function 2^n is unary representation, and
  (2) $r(n) = O(n^*)$, $s(n) = O(n^*)$

17.3 Let M be a Turing machine with only one work tape. Prove that
$$s(n) = O(n \cdot r(n))$$

17.4* Prove that if $s(n) = o(\log\log n)$, then $s(n) = O(1)$.

§18 LOG-SPACE TRANSFORM MACHINE

As a tool, we propose a computational model LSTM: log-space transform machine.

DEFINITION 18.1 A transform machine is a Turing machine with a special state q. The input alphabet and output alphabet are the same. It works like an ordinary Turing machine. But when it enters the special state q, it removes all the contents from the input tape, transfers the whole contents of the output tape to the input tape, changes the work tapes and the output tape to blanks, moves its input head to the leftmost symbol, then enters the initial state q_0 and works normally. The machine halts when it goes into a terminating state. Then the contents of the output tape are considered to be the result of the computation.

The space complexity $s(n)$ of a transform machine is defined as that of Turing machines. The width complexity $w(n)$ is the maximum total length of the input and output tape contents during the computation for all inputs of length $\leq n$. The phase complexity $p(n)$ is one plus the total number of times of its entering q during the computation for all inputs of length $\leq n$.

DEFINITION 18.2 If a transform machine satisfies

$$s(n) = O(\log w(n)),$$

then we call it a log-space transform machine, or simply an LSTM.

EXAMPLE 18.1 There is an LSTM satisfying the conditions that
   INPUT: $x_1\#x_2\#\ldots\#x_k{}^*y_1\#y_2\#\ldots\#y_m$ where $x_i, y_i \in (0+1)^*$.
   OUTPUT: $x_1y_1\#x_1y_2\#\ldots\#x_1y_m\#x_2y_2\#\ldots\#x_2y_m\#\ldots\#x_ky_1\#x_ky_2\#\ldots\#x_ky_m$.
   COMPLEXITY: $p(n) = 1$, $w(n) = O(n^2)$.

In fact, all we need are two pointers i and j indicating that x_iy_j is now copied onto the output tape.

EXAMPLE 18.2 There is an LSTM M satisfying the conditions that
   INPUT: $x_1\#x_2\#\ldots\#x_k$ where x_i's are binary natural numbers
   OUTPUT: Σx_i in binary.
   COMPLEXITY: $p(n) = O(\log n)$, $w(n) = O(n)$.

Proof: in the first phase, M gets $y_1\#y_2\#\ldots$ on its output tape, where $y_i = x_{2i} + x_{2i+1}$.
   In the second phase, M gets $z_1\#z_2\#\ldots$ on its output tape, where

$z_i = y_{2i} + y_{2i+1}$.

Thus in $O(\log k)$ phases, M gets Σx_i on its output tape. The width is obviously $O(n)$. Because of $k \leq n$, $p(n) = O(\log n)$.

DEFINITION 18.3 A class of problems is in NC, if there exists an LSTM solving it in phase $O(\log^* n)$ and width $O(n^*)$.

Nick Pippenger made the first study of NC by means of TM of $\log^* n$ reversal and n^* space. The above definition is different from the original one, but we will prove their equivalence later.

EXAMPLE 18.3 Addition, subtraction, multiplication and division of binary natural numbers belong to NC. The proofs are left as exercises.

THEOREM 18.1 Suppose that M is an LSTM with phase $p(n)$ and width $w(n)$. Then there exists an LSTM M' satisfying the conditions that

INPUT: $x_1 \# x_2 \# \ldots \# x_k$ where # is a separator.

OUTPUT: $M(x_1) \# M(x_2) \# \ldots \# M(x_k)$.

COMPLEXITY: phase $p'(n) = O(p(n))$, width $w'(n) = O(n \cdot w(n))$.

Proof: in one phase, M' simulates one phase of M for x_1, writes the result on output tape and adds a separator #, then for x_2, then for x_3,.... So in one phase, M' gets the one-phase results of M for all the x_i's on the output tape. Then M' enters its special state and begins the next phase in the same way.

Obviously, in $\text{Max}\{p(|x_i|) \mid i = 1,2,\ldots,k\}$ phases, M' will get

$M(x_1) \# M(x_2) \# \ldots \# M(x_k)$

on its output tape. So

$p'(n) \leq \text{Max}\{p(|x_i|) \mid i = 1,2,\ldots,k\} \leq p(n)$

$w'(n) \leq 2k + w(|x_1|) + w(|x_2|) + \ldots + w(|x_k|)$

$\leq 2k + k \cdot w(n)$

$\leq 2n + n \cdot w(n)$

$= O(n \cdot w(n))$

$$s'(n) \leq \text{Max } \{\log k + s(|x_i|)|i = 1,\ldots,k\} = O(\log w(n)) = O(\log w'(n))$$

THEOREM 18.2 Suppose that L_1 and L_2 are two LSTM's of phase one, and that the length of w is bounded by a polynomial of the length of $L_2(w)$ (or $|L_1(w)|$ is bounded by a polynomial of $|w|$). Then $L_1 \cdot L_2$ is an LSTM of phase one.

Proof: we use the same method as in the proof of Theorem 17.4. For input w of length n the space needed to simulate L_1 is $O(\log (n+|L_1(w)|))$. The space needed for the counter is $O(\log|L_1(w)|)$. The space needed to simulate L_2 is $O(\log(|L_1(w)| + |L_2(L_1(w))|))$. Therefore the total space needed is $O(\log(|L_1(w)| + n + |L_2(L_1(w))|)) = O(\log(n+|L_2(L_1(w))|))$. Thus there is an LSTM L of phase one simulating $L_1 \cdot L_2$.

Exercises

18.1 Prove Example 18.3.

18.2 Binary multiplication can be computed by an LSTM of phase 1.

18.3* Suppose that x is a binary number of length n. Then the binary number $[\sqrt{x}]$ can be computed by an LSTM of phase $\log^* n$ and width n^*.

§19 LOG-SPACE CONSTRUCTIBLE GRAPHS AND NICE PAIR OF FUNCTIONS

Later in this book, we mainly discuss the upper bounds of complexities. Usually we choose some simple and easily computable functions as these bounds, such as n^2, $n \cdot \log n$, 2^n, etc. All the functions below refer to functions from natural number set to natural number set. Where the value of the function is not an integer, we really mean its integer part. For example, by $n \cdot \log n$ we mean $[n \cdot \log n]$.

DEFINITION 19.1 Function $f(n)$ is called log-space constructible, if there exists an LSTM of phase 1 transforming n to $f(n)$ in unary.

A family of codings $\{C_n|n = 1,2,\ldots\}$ is called log-space constructible, if there is an LSTM of phase 1 transforming a^n to C_n (a is a letter).

EXAMPLE 19.1 n, log n, 2^n are all log-space constructible.

Proof: obvious.

EXAMPLE 19.2 If $f(n)$ is log-space constructible, then so is $f(n)^2$.

Proof: suppose L_1 is an LSTM of phase 1 computing $f(n)$ and L_2 computing n^2 in unary form with

$$s_1(n) = O(\log(n+f(n)))$$

$$s_2(n) = O(\log n).$$

Then $L = L_1 \cdot L_2$ is the desired LSTM (see Theorem 18.2).

For the same reason, $f(n)^k$ is log-space constructible for any positive integer k.

Suppose that $G = (V,E)$ is a labelled directed graph with vertex set V and edge set E. We name the vertices in G by integers $1,2,\ldots,n = |V|$. For each vertex v, if it is labelled with a, its fan-in number is k and (v_1,v), $(v_2,v),\ldots,(v_k,v)$ are the k edges into it, then we use the following information segment as its coding (the letters a, v,... are all binary strings):

$$(a,v,v_1,v_2,\ldots,v_k).$$

The coding of the graph is the concatenation of information segments of all vertices in some order. Since there are many ways to name the vertices and the order of information segments may vary, there are many different codings for the same graph.

DEFINITION 19.2 A family of graphs $\{G_i | i = 1,2,\ldots\}$ is log-space constructible if a family of its codings is log-space constructible.

EXAMPLE 19.3 (The FFT network) The graph FFT_ℓ consists of the vertices $\{(i,j) | i = 0,1,\ldots,\ell,\ 0 \leq j \leq 2^\ell - 1\}$. We represent j as a binary string of length ℓ, $j:j_0 j_1 \ldots j_{\ell-1}$. The edge set of FFT is

$$\{((i,j_0 j_1 \ldots j_i \ldots j_{\ell-1}), (i+1, j_0 j_1 \ldots 0 \ldots j_{\ell-1})),$$
$$((i,j_0 j_1 \ldots j_i \ldots j_{\ell-1}), (i+1, j_0 j_1 \ldots 1 \ldots j_{\ell-1})) | i = 0,1,\ldots,\ell-1,$$
$$j_0, j_1, \ldots, j_\ell = 0,1\}$$

For $\ell = 3$, the FFT_3 is shown by Fig. 19.1.

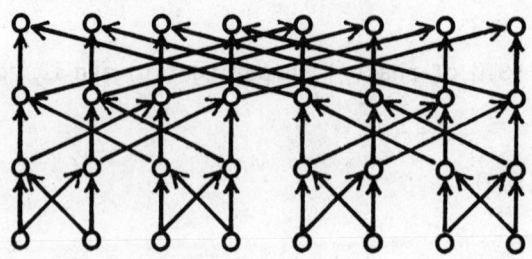

Fig. 19.1

We can use some counters of length ℓ (= $O(\log L)$, where L is the length of the coding of the graph) to generate the coding. Therefore we can transform 1^ℓ to the coding of FFT_ℓ by an LSTM of phase one. Notice that the coding is not in the form of our definition, but we can transform it into the form of our definition by an LSTM of phase one. By Theorem 18.2, we know that the family of FFT graphs is log-space constructible.

Suppose that there are two sets U and V. A relation from U to V is a subset of U × V. For each element (u,v) in the set U × V, we have a coding (C_u, C_v), where C_u and C_v are the names (binary strings) of u and v respectively. The coding of a relation $R \subseteq U \times V$ is the concatenation of the codings of all elements of R in some order.

DEFINITION 19.3 A family of relations $\{R_i \subseteq U_i \times V_i | i = 1,2,...\}$ is log-space constructible if a family of their codings is log-space constructible.

DEFINITION 19.4 Suppose that $G' = (V',E')$ and $G'' = (V'',E'')$ are graphs, $R' \subseteq V' \times V''$ and $R'' \subseteq V'' \times V'$ are relations. A graph G is called the graph composed from G' and G'' by relations R' and R'' if G = (V,E), V = V' ∪ V'' and E = E' ∪ E'' ∪ R' ∪ R'', where V' and V'' are considered to be disjoint sets.

EXAMPLE 19.4 Let G' and G'' be the graphs shown by Fig. 19.2. The relations R' and R'' are

$$R' = \{(1,1), (2,2), (3,3)\}, \quad R'' = \{(1,3), (3,1)\}.$$

Then the graph G composed from G' and G" by R' and R" is as shown in Fig. 19.3.

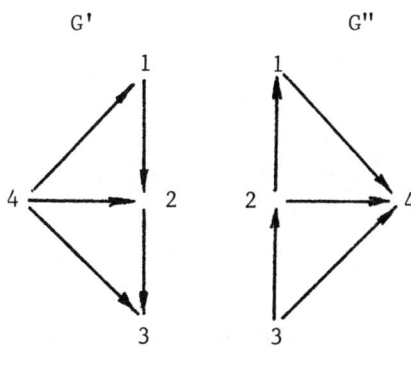
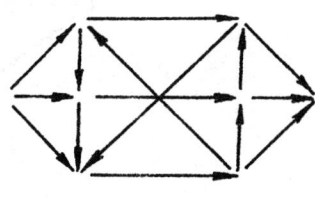

Fig. 19.2　　　　　　　　　　　　　　　Fig. 19.3

THEOREM 19.1　If the families of graphs $\{G_i' | i = 1,2,...\}$ and $\{G_i'' | i = 1,2,...\}$, the families of relations $\{R_i' | i = 1,2,...\}$ and $\{R_i'' | i = 1,2,...\}$ are all log-space constructible, then the family of graphs $\{G_i | i = 1,2,...\}$, composed from G_i' and G_i'' by R_i' and R_i'', is log-space constructible.

Proof: we transform a^i to the codings of G_i', G_i'', R_i', R_i'' (concatenated together) by an LSTM of phase 1. Given these codings as input, we can obtain the coding of G_i by an LSTM of phase one. By Theorem 18.2 we know that there is an LSTM of phase one transforming a^i to the coding of G_i. Thus the family of G_i is log-space constructible.

DEFINITION 19.5　Suppose that $G' = \{V',E'\}$ and $G'' = \{V'',E''\}$ are graphs with $V' \cap V'' = \phi$, the relation $R \subseteq V' \times V''$ is a one to one correspondence between a subset $S' \subseteq V'$ and $S'' \subseteq V''$. The graph G is called the join of G' and G" by R, if G is obtained by identifying the corresponding vertices in S' and S" according to R.

EXAMPLE 19.5 We join the graphs shown in Fig. 19.2 by the relation

$$R = \{(1,1),(2,2),(3,3)\}.$$

The result is shown in Fig. 19.4

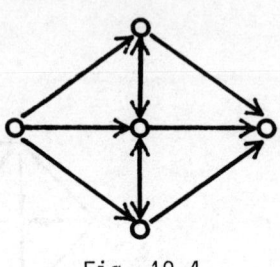

Fig. 19.4

THEOREM 19.2 If $\{G_i'|i = 1,2,\ldots\}$, $\{G_i''|i = 1,2,\ldots\}$ and $\{R_i|i = 1,2,\ldots\}$ are all log-space constructible, then so is the family of graphs $\{G_i|i=1,2,\ldots\}$ obtained by joining G_i' and G_i'' according to relation R_i.

Proof: use the same technique as in the proof of Theorem 19.1.

DEFINITION 19.6 Let $G = (V,E)$ be a graph, S', $S'' \subseteq V$ be two subsets of vertices with $S' \cap S'' = \emptyset$, C be a one-to-one correspondence from S' to S'' (therefore $|S'| = |S''|$). Suppose that $G_i = (V_i,E_i)$, $i = 1,2,\ldots,m$, are all isomorphic to G, and ϕ_i is an isomorphic from G_i to G_{i+1}, $\phi_i : V_i \to V_{i+1}$ such that the corresponding subsets S_i', $S_i'' \subseteq V_i$ are mapped to S'_{i+1} and S''_{i+1} respectively by ϕ_i. Let C_i be the correspondence from S'_i to S''_i induced from C by ϕ_i. For any vertex $v \in S_i'$, identifying vertex $C_i(v) \in S_i''$ with vertex $\phi_i(v) \in S'_{i+1}$, $i = 1,2,\ldots,m-1$, we obtain a simple graph. This graph is called the m-multilevelled graph of G induced by C.

EXAMPLE 19.6 Let G be the left graph in Fig. 19.5, $S' = \{1,2\}$, $S'' = \{3,4\}$, and C be $\{1 \to 3, 2 \to 4\}$. Then the 3-multilevelled graph of G induced by C is the right-hand graph in Fig. 19.5.

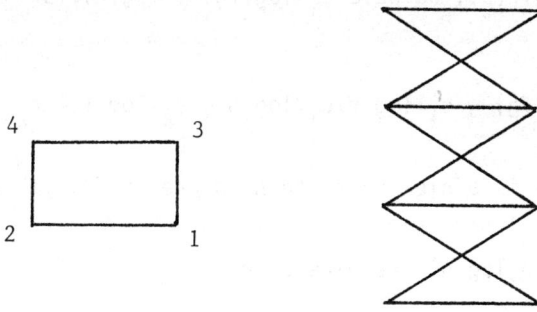

Fig. 19.5

THEOREM 19.3 Suppose that the family of graph $\{G_i | i = 1,2,...\}$, the family of correspondences $\{C_i | i = 1,2,...\}$ and the function $h(n)$ are all log-space constructible. Then the family of $h(i)$-multilevelled graphs of G_i induced by $C_i (i = 1,2,...)$ are log-space constructible.

Proof: use the same technique as above. The details are left as an exercise.

DEFINITION 19.7 Let $(f(n), g(n))$ be a pair of functions. If

(1) both $f(n)$ and $g(n)$ are non-decreasing and log-space constructible,

(2) either $f^*(n) \geq g(n)$ or $g^*(n) \geq f(n)$,

(3) $(f(n) \cdot g(n))^* \geq n$ and

(4) $f(n) \leq 2^{g^*(n)}$, $g(n) \leq 2^{f^*(n)}$,

then we call $(f(n), g(n))$ a log-space constructible nice pair of functions with respect to reversal and space, or, briefly, a nice pair (of functions).

In the above definition, (1), (2) are for the sake of simplicity of f and g, (3) is for the sake of Theorem 17.5, (4) is for the sake of 17.4, 17.5, 17.7 and 17.8.

LEMMA 19.1 If (f,g) is a nice pair, then $f^*(n) \geq \log n$, $g^*(n) \geq \log n$.

Proof: without loss of generality, we can assume $f^* \geq g$ and prove $g^* \geq \log n$. By (4), we have $g^* \geq \log f$. Since $f^* \geq g$, we have $c \cdot \log f \geq \log g$ for some constant c. Because of (3), we have $c_1 \cdot \log(fg) \geq \log(n)$ for some constant c_1. Therefore

$$\log n \leq c_1 \cdot \log(fg) \leq c_1 \cdot \log f + c_1 \cdot \log g \leq c_2 \cdot \log f \leq c_2 \cdot g^* \leq g^*.$$

EXAMPLE 19.7 If (f,g) is a nice pair, then so are (g,f), (f^2,g^2), (f^3,g^4).

EXAMPLE 19.8 (n,n), $(n,\log n)$ are nice pairs.

$(2^n, \log n)$ is not a nice pair because of (4).

$(\log\log n, \log n)$ is not a nice pair because of (3).

LEMMA 19.2 If (f,g) is a nice pair, then either $f^*(n) \geq n$ or $g^*(n) \geq n$.

Proof: if $f^* \geq g$, then $n \leq (f \cdot g)^* \leq (f \cdot f^*)^* \leq (f^*)^* \leq f^*$. The same for $g^* \geq f$.

If the reversal complexity and space complexity of some Turing machine M are bounded by a nice pair $(f(n), g(n))$:

$r(n) = O(f(n))$

$s(n) = O(g(n))$,

then we have

$t(n) = O(n \cdot r(n) \cdot s(n))$

$\quad\quad = O((f(n) \cdot g(n))^* \cdot f(n) \cdot g(n))$

$\quad\quad = O((f(n) \cdot g(n))^*)$.

So we can express the three resources of a Turing machine M as a rectangle: the base indicates the space complexity $g(n)$; the height indicates the reversal complexity $f(n)$; the area is the time complexity $(f(n) \cdot g(n))^*$. This situation is shown in Fig. 19.6 and Fig. 19.7, here we do not distinguish the functions that are polynomially related.

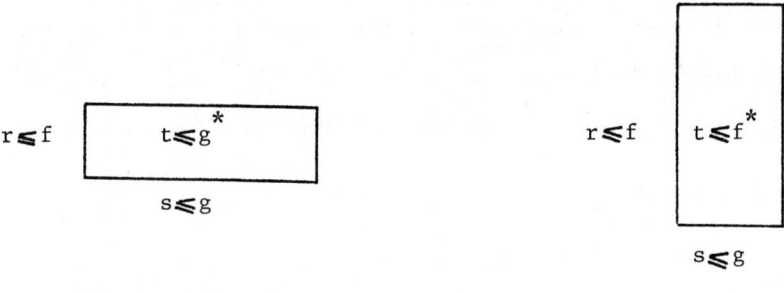

Fig. 19.6   Fig. 19.7

This rectangle may be flat ($g^* \geq f$) or narrow ($f^* \geq g$).

In the flat case, the time complexity is essentially the same as the space complexity:

$$t(n) = O((f(n) \cdot g(n))^*) = O(g^*(n)).$$

In the narrow case, the time complexity is essentially the same as the reversal complexity:

$$t(n) = O((f(n) \cdot g(n))^*) = O(f^*(n)).$$

Exercises (prove the following statements)

19.1 If f and g are log-space constructible, then so are

    (1) $f + g$

    (2) $f \cdot g$

    (3) 2^f

    (4) f^g

19.2 $(n, 2^n)$, (n, n^n) are nice pairs.

19.3 If (f, f) is a nice pair, then so is $(f, 2^f)$.

19.4 If (f, g) is a nice pair, then so is (f^*, g^*).

19.5 f(n) is called a nice function, if

(1) f(n) is non-decreasing and log-space constructible,

(2) $f^*(n) \geq \log n$.

Prove that f is a nice function iff there exists a function g such that (f,g) is a nice pair.

19.6 If f(n) and g(n) are both log-space constructible and

$f^*(n) \geq n$ or $g(n) \leq n^*$

then f(g(n)) is log-space constructible.

§20 THE CONCEPT OF SIMILARITY

Because of the stimulus of parallel computation and complexity theory, people have proposed many computational models. For example, multitape Turing machines, RAM (Aho, Hopcroft, Ullman, 1974), vector machines (Pratt and Stockmeyer, 1978), storage modification machines (Schönhage, 1979), hardware modification machines (Dymond and Cook, 1980), uniform circuits (Borodin, 1977), uniform aggregates (Dymond and Cook, 1980), VLSI (Brent and Kung, 1980; Thompson, 1979), and so on. Each model has several different resources. Even for one model, there are many different computational types (e.g., deterministic, non-deterministic, alternating,...). Therefore one can ask whether the complexity for a given problem class is a model-independent objective reality, how to unify them, etc.

In recent years, people have begun to notice relations between different models. Cook and Reckhow (1973) and many others noticed that not only can different computational models simulate each other in principle, but also their sequential time complexities are polynomially related, hence strengthening the Church-Turing thesis. Pratt and Stockmeyer (1978) gave the first example of the polynomial equivalence of parallel time and sequential space. Chandra and Stockmeyer (1976) and Goldschlager (1978) proposed independently a parallel computation thesis which says that, for parallel computational models, parallel time corresponds to space. Borodin (1977) pointed out that deterministic Turing machine time corresponds to uniform circuit size, while deterministic Turing machine space corresponds to uniform circuit depth. He also pointed out that in fact the equivalence of Turing machine time and uniform circuit size goes back to works of Schnorr, Pippenger and Fischer. Pippenger (1979) proved that Turing machine time and space correspond to uniform circuit size and width simultaneously, while Turing machine time and reversal correspond to uniform circuit size and depth simultaneously. Dymond (1980) proved that Turing machine reversal and space correspond to uniform circuit depth and width simultaneously. He also proposed a extended parallel computation thesis that Turing machine reversal and space correspond to parallel time and hardware. Later, Dymond and Cook (1980) proposed uniform aggregates, and proved this thesis for it.

Despite their number and variety, the basic resources remain:

parallel time (e.g., the reversal of a Turing machine)

space (e.g., the space of a Turing machine)

sequential time (e.g., the time of a Turing machine).

The author (1980) proposed the similarity principle: all reasonable and strong enough computational models can not only simulate one another, but also simultaneously use essentially the same parallel time, essentially the same space and essentially the same sequential time while simulating one another. Here "essentially the same" means "polynomially related". He proved the similarity principle for twelve main computational models.

Notice that the above three resources have different names and different definitions in different models. What we keep in mind as a principle is:

(1) The parallel time is the number of phases, while a phase is a period of computation during which no intermediate computation results written down within this period are used.

(2) The space is the maximum length of the intermediate computation results that have to be kept for later use.

(3) The sequential time is the total number of primitive operations.

For most computational models, sequential time does not exceed a polynomial of the product of parallel time and space. Later on, we mainly discuss parallel time and space.

Because of the different names of the same resource in different models, we must specify the name of parallel time and space in the model discussed. If MODEL is a computational model, in which the resources corresponding to parallel time and space are called R and S respectively, then we denote it as

    MODEL (R,S)

and the corresponding complexity functions are denoted by R(n) and S(n). For example, the Turing machine will be denoted as

    TM (reversal, space),

while reversal(n) and space(n) are the reversal complexity and the space complexity respectively, i.e., r(n) and s(n) in the previous part of this chapter.

Let $(f(n), g(n))$ be a nice pair and M be a machine in MODEL(R,S). If

    $R(n) \leq f(n)$

    $S(n) \leq g(n)$,

then we say that M is an (f,g)-machine in MODEL(R,S).

If for any (f,g)-machine M in MODEL(R,S) there exists an (f^*,g^*)-machine M' in MODEL'(R',S') simulating M, then we say that MODEL(R,S) can be polynomially simulated by MODEL'(R',S') and denote it by

    MODEL(R,S) $\leq$ *MODEL'(R',S').

If both

    MODEL(R,S) $\leq$ *MODEL'(R',S').

and

    MODEL'(R',S') $\leq$ *MODEL(R,S)

hold, then we say that MODEL(R,S) and MODEL'(R',S') are similar and write

    MODEL(R,S) = *MODEL'(R',S').

In this book, we will prove the similarity of

    TM(reversal, space)
    LSTM(phase, width)
    RAM(reversal, space)

VM(time, space)
UC(depth, width)
UA(time, space)
PRAM(time, space)

The first three are sequential computational models, the others are parallel ones. We mainly discuss the flat case, that is, $f \leq g^*$. The proofs in the narrow case are trivial and are left to the reader.

REMARK 20.1 In the flat case, we have

n = length of input $\leq g^*(n)$

length of output $\leq g^*(n)$.

So all the input and output can be stored in workspace. While defining RAM and VM, we do not separate the input and output space from the work space. If one wants to discuss the narrow case, the separation of the input and output space from the work space is necessary, because, for example, $g(n)$ may be log n which is much less than the input length.

REMARK 20.2 We shall see later that the input alphabet of any VM, UC and UA is {0,1}. In order to standardize the unit of the input length while discussing simulation between different models, from now on we take the input alphabet of any machine to be {0,1}. Later, sometimes say that, for example, the input of some machine is aaaa#bb*abab. Actually the symbols a,b #, * are abbreviations of some {0,1}-coding, say, 00,01,10,11 respectively.

Exercise

20.1 Prove that the single work tape Turing machine and the multitape TM are not similar.

5 Multi-index random access machine

§21 DEFINITION

A random access machine (or simply, RAM) is a very important and widely used computational model. It is quite similar to modern electronic computers. To translate an RAM program to a program of a realistic computer is straightforward.

A multi-index random access machine is a precise variety of RAM proposed in [Cook & Reckhow, 1973]. Its only difference is that it has some additional index registers which make the discussion of reversal in RAM possible. In this book, RAM will refer to the multi-index random access machine.

In this chapter, we deal only with multi-index RAM in the flat case, i.e. $g^* \geq f$. In this case we have $g^* \geq n$, therefore we do not distinguish the input and output space from the work space.

An RAM consists of the following three parts (Fig. 21.1).

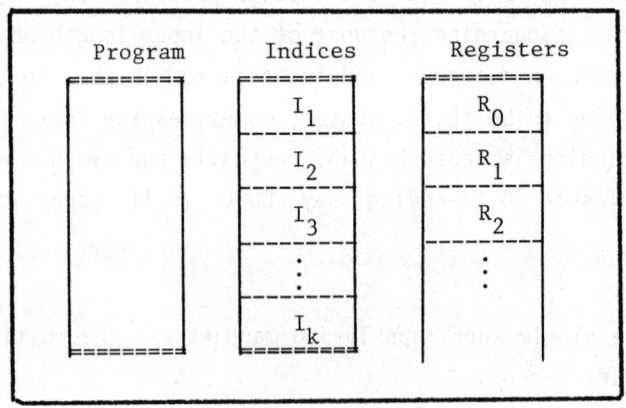

Fig. 21.1 RAM

(1) Infinitely many (ordinary) registers

$$R_0, R_1, R_2, R_3, \ldots$$

each of which has the ability to store an arbitrary natural number, no matter how large it is. The integer stored in R_i is denoted by

   Content(R_i).

(2) k index registers, or indices

   $I_1, I_2, I_3, \ldots, I_k$.

An index register is a register for indirect addressing (see (3)). Each index register also has the ability to store an arbitrary natural number. The integer stored in I_i is denoted by

   Content(I_i).

(3) A program, described as follows.

   (A) Address:
      direct address: I_j or R_j.
      indirect address: #I_j, indicating R_m where m = Content(I_j).

   (B) Instruction: There are three kinds of instructions as follows.

      A:=a, meaning to place the natural number a in A;
      A:=B, meaning to place the number Content(B) in A;
      A:=B*C, meaning to place the number Content(B)*Content(C) in A

where A,B,C are direct or indirect addresses and * is any mapping in NC. We call such mappings admissible instructions. For example addition "+", subtraction "-", multiplication "*" and integer division "/", are all admissible.

   (C) The program is a directed graph. The vertices in this graph are called states of this program. The fan-out of every vertex in this graph is less than or equal to 2. If the fan-out of a vertex is 1, then the unique edge out from this vertex is labelled with an admissible instruction; if the fan-out of a vertex is 2, then these two edges out from this vertex are labelled with two opposite predicates $I_j = 0$ and $I_j \neq 0$ (or $R_j = 0$ and $R_j \neq 0$); if the fan-out of a vertex is 0, then this vertex is called a terminating state. A specified state q_0 is called the initial state of this program.

The input is always a binary string of length n : $a_1 a_2 \ldots a_n$. There are three different input forms. In the first form, the i-th input bit a_i is stored in R_i (as an integer 0 or 1) the integer n is stored in R_0 and R_{n+1}.

In the second form, the integer

$$\sum_{i=1}^{n} a_i \, 2^{i-1}$$

is stored in R_1, the integer 1 is stored in R_0 and the length n is stored in R_2. In the third form, $n = m\ell$, the integers

$$\sum_{i=1+mj}^{m(j+1)} a_i \, 2^{i-1-mj} \qquad j = 0,1,2,\ldots,\ell-1$$

are stored in R_1, R_2, R_3,...,R_ℓ respectively and the integers ℓ and n are stored in R_0 and $R_{\ell+1}$ respectively. For any form, the input length is n. In fact, the first two forms are special cases of the third form.

Notice that in the second form if content(R_2) = 3, content(R_1) = 5, then the input is a binary string 101. But if content(R_2) = 5, content(R_1) = 5, then the input is a binary string 00101.

The RAM starts from q_0. Whenever in a state of fan-out 1, it performs the corresponding admissible instruction and enters the next state. Whenever in a state of fan-out 2, it takes the branch on which the corresponding predicate is satisfied. If it enters a state of fan-out 0, it halts and the contents in some designated registers are the output. For convenience, where there is no confusion, we will write R_i instead of content(R_i). For example, we may write $R_i = 9$, $I_i = 1,2,\ldots$ and so on.

EXAMPLE 21.1 In Fig. 21.2, we give a program which decides whether the number of 0's is equal to the number of 1's in the input {0,1}-string:

INPUT : R_i = 0 or 1 (i=1,2,...,n), $R_0 = R_{n+1} = n$.

OUTPUT: R_0 = 1 or 0 (represents YES or NO)

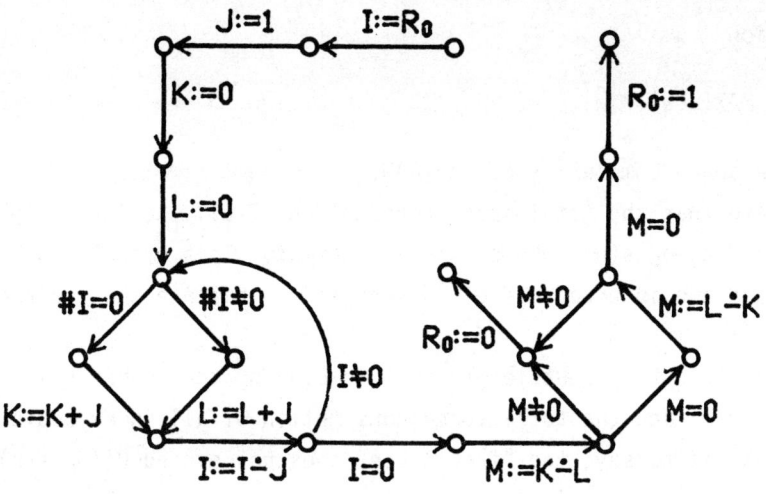

Fig. 21.2

This program can be expressed as

```
    BEGIN
        I:=R_0; J:=1; K:=0; L:=0;
*:      IF #I=0 THEN K:=K+J ELSE L:=L+J;
        I:=I-J:
        IF I≠0 THEN GOTO * ELSE
            IF K=L THEN R_0:=1 ELSE R_0:=0;
    END
```

where the I,J,K,L are all indices.

About the complexities of an RAM, we give the following explanation:

1. Space

The space consumption of a register R (or an index register) is the binary length of the maximal integer stored in it during the computation, denoted by Cost(R).

For an input w (a binary string), the sum of the space consumption of the ordinary registers that are used during the computation is called the space consumption of the RAM for w, denoted by s(w).

The function

$$s(n) = \text{MAX}\{s(w) \mid |w| \leq n\}$$

is called the space complexity of the RAM.

We stipulate that the total space consumption of all the index registers does not exceed $c \cdot \log s(n)$, where c is a constant. This stipulation is natural because the intended use of indices is to indicate the indirect addresses.

In Example 21.1, the input length is n (excluding the number n in R_0 and R_{n+1}), $s(n) = O(n)$ and the total space consumption of all index registers is $O(\log n)$. That is to say, the "free space" does not exceed $O(\log s(n))$.

Here all the work space (ordinary registers) is used for the storage of the input and all the "computation" is performed in index registers, whose space consumption is not counted in the space complexity. This is a good illustration of the power of indices. The positions of tape heads in TM correspond to the indices in RAM. But the former only indicate the "address", while the latter, although restrained to $O(\log s(n))$, can do all the operations, so are much more powerful.

2. Reversal

A phase is a period of computation within which no ordinary register is first written in and then read out. For an input w, the minimal number of the phases covering the whole computation is called the reversal for w, denoted by r(w). The function

$$r(n) = \text{MAX}\{r(w) \mid |w| \leq n\}$$

is called the reversal complexity.

It is just as important that the reading and writing of indices do not interrupt a phase, as that change of the positions of tapeheads along one direction in a Turing machine does not interrupt a phase. This is the essential difference between ordinary RAM and multi-index RAM.

In Example 21.1, $r(n) = 1$. That is to say, RAM can accept the language

$$L = \{w \in (0+1)^* \mid \text{the numbers of 0's and 1's in w are equal}\}$$

in one phase, while the Turing machine cannot. Again we see how powerful the indices are.

3. **Time** (sequential time)

The time consumption of various instructions is defined in Table 21.1.

INSTRUCTION	TIME CONSUMPTION
A := a	$|a|$, the length of integer a
A := B	$|Content(B)|$
A := B*C	$|Content(B)| + |Content(C)|$

Table 21.1

The time consumption of checking predicate $I = 0$ and $I \neq 0$ (or $R_i = 0$, $R_i \neq 0$) is $|Content(I)|$ (or $|content(R_i)|$).

For an input w, the sum of the time consumptions of executing all instructions and checking all predicates is called the time consumption for w, denoted by $t(w)$. The function

$$t(n) = MAX\{t(w) \mid |w| \leq n\}$$

is called the time complexity.

As in the TM case, we discuss only RAM's with $t(n) < \infty$.

The description (or simply, ID) of an RAM at any given moment is

$$[q(1,i_1)(2,i_2)...(k,i_k)](0,c_0)(1,c_1)(2,c_2)...(m,c_m)$$

indicating that the state of the RAM is q, $Content(I_j)$ is i_j ($j = 1,2,...,k$) and $Content(R_j)$ is c_j ($j = 0,1,2,...,m$) where $R_0, R_1, R_2,...,R_m$ are all the ordinary registers that are used up to now. The first part $[q(1,i_1)(2,i_2)...(k,i_k)]$ is called the prefix of the ID, the remainder $(0,c_0)(1,c_1)(2,c_2)...(m,c_m)$ is called the suffix of the ID. (It is not necessarily consecutive. For example, if R_3 is not used then $(3,c_3)$ does not appear in it.)

Obviously, the instruction that will be executed and the next state are uniquely determined by the current ID.

It is not difficult to prove that within a phase, the prefixes of any two ID's are not equal. Otherwise the machine would fall into an endless cycle. Because the "free space" is $O(\log s(n))$, there are at most $s^*(n)$ different prefixes. So any phase can cover a time period of length at most $s^*(n)$. Thus, for any RAM, we have

<u>THEOREM 21.1</u> $t(n) = O(r(n) \cdot s^*(n))$

The consumptions defined above are called the logarithmic costs (because the space cost of an integer is approximately the logarithm of the integer and the same for sequential time) and are mainly used in this book. We can also define the uniform costs for space and time (the reversal remains the same) if we define the uniform space cost for input w to be the total number of ordinary registers used and uniform time cost to be the total number of instructions executed and predicates checked.

We can further restrict the RAM model so that:

(1) the admissible instructions are only arithmetic operations between natural numbers, i.e., addition +, subtraction $\dot{-}$, multiplication * and integer division /, where subtraction $\dot{-}$ is defined as

$\quad$ a $\dot{-}$ b = if a $\geq$ b then a-b else 0;

(2) the (logarithmic) space consumption of any register is $O(g^*(n))$, where $g(n)$ is the uniform space complexity, i.e., the total number of ordinary registers used.

Therefore if the input is in the second or third form, then the length of each input integer must be bounded by $O(g^*(n))$.

This restricted model will be called RAM_0.

In the case of RAM_0, the logarithmic costs are polynomially related to uniform costs. We have the following theorem.

<u>THEOREM 21.2</u> If the uniform space and time complexities of an RAM_0 are $O(g(n))$ and $O(t(n))$ respectively, then the logarithmic space and time complexities are $O(g^*)$ and $O(t^*)$ respectively.

Proof: by definition, a register is of length at most $O(g^*)$ therefore the logarithmic space cost is $O(g \cdot g^*) = O(g^*)$ and the logarithmic sequential time cost is $O(t \cdot g^*) = O(t^*)$.

Exercises

21.1 Design an RAM to calculate the GCD (greatest common divisor) of two positive integers:

INPUT: $R_0 = m$, $R_1 = \ell$ where m and ℓ are positive integers (notice that the input is not in one of our input forms in the text)

OUTPUT: $R_2 = GCD(m, \ell)$

COMPLEXITY: $r(n) = O(n)$, $s(n) = O(n)$, where n is the input length and the logarithmic costs are used.

21.2 Design an RAM to calculate the Fibonacci sequence:

INPUT: $R_0 = m$

OUTPUT: $R_0 = f(m)$ where $f(0) = 1$,
$f(1) = 1$,
$f(m) = f(m-1) + f(m-2)$ for $m \geq 2$.

COMPLEXITY: $r(n) = O(2^n)$, $s(n) = O(2^n)$, where n is the binary length of m.

21.3 Design an RAM_0 to accept the following language:

(1) $\{0^n 1^n 0^n; n = 0, 1, \ldots\}$

(2) $\{ww; w \in (0+1)^*\}$

What is the complexity of your program?

21.4 Design an RAM to transform input of the second form to input of the first form.

21.5 Suppose that i is the largest number such that R_i is used. Prove that $\log i = O(\log s(n))$. Therefore $i = O(s^*(n))$.

§22 EXAMPLES

EXAMPLE 22.1 (Max)

INPUT: $R_0 = \ell$, $R_1 = a_1$, $R_2 = a_2$,...,$R_\ell = a_\ell$, $R_{\ell+1} = n$; where a_i are positive integers, n is total input length.

OUTPUT: $R_0 = \text{Max}\{a_i | 1 \leq i \leq \ell\}$

PROGRAM: $I:=R_0$; $R_0:=0$;

$*:R_{\ell+1}:=\#I-R_0$; IF $R_{\ell+1} \neq 0$ THEN $R_0 := \#I$; $I:=I-1$; IF $I=0$ THEN stop; GOTO$*$.

COMPLEXITY: $r(n) = O(n)$, $s(n) = O(n)$ (logarithmic cost).

The registers R_0 and $R_{\ell+1}$ are written and read alternately, therefore the reversal is $O(\ell) = O(n)$. It is not difficult to design an RAM of reversal $O(\log n)$ and space $O(n)$ to do the work. But the program will be much more complicated and is left to the reader.

The next example together with Exercise 21.4 shows that all input forms can easily be transformed to each other.

EXAMPLE 22.2 (Transform input of the first form to input of the second form).

INPUT: $R_0 = n$, $R_1 = a_1$, $R_2 = a_2$,...,$R_n = a_n$, $R_{n+1} = n$.

OUTPUT: $R_0 = 1$, $R_1 = a_n 2^{n-1} + a_{n-1} 2^{n-2} + \ldots + a_1$, $R_2 = n$.

COMPLEXITY: $O(\log n)$ reversal, $O(n \log n)$ space (logarithmic cost).

Without loss of generality, we assume that n is a power of 2. We compute $a_1 + 2a_2$, $a_3 + 2a_4$, $a_5 + 2a_6$,... into R_2, R_4,... respectively. Then compute $(a_1 + 2a_2) + 2^2(a_3 + 2a_4)$, $(a_5 + 2a_6) + 2^2(a_7 + 2a_8)$,... into R_4, R_8,... respectively, and so on.

```
BEGIN
    I := 1;  R_0:=2;
    WHILE I ≤ n/2 DO
        BEGIN I_1:=I;  I_2:=I_1+1;
            WHILE I_1 ≤ n DO
                (#I_2:=#I_1+#I_2*R_0;  I_1:=I_1+2*I ;  I_2:=I_2+2*I );
            I :=2*I ;  R_0:=R_0*R_0;
        END
    R_1:=R_n
END.
```

There are all together $O(\log n)$ phases. In each phase, the new space used is $O(n)$. Therefore the total space used is $O(n \log n)$.

EXAMPLE 22.3 Shortest-path problem

INPUT: an $m \times m$ 0-1 matrix, which is the adjacency matrix of a graph.

$R_1 = a_{11}$, $R_2 = a_{12}$, ..., $R_m = a_{1m}$, $R_{m+1} = a_{21}$, ..., $R_{m^2} = a_{mm}$.

$R_0 = R_{m^2+1} = n = m^2$.

OUTPUT: the length ℓ of the shortest path between vertex i and m.

We use the well-known Breadth First Search technique. At first no vertices in the graph are considered to be labelled.

1. Label vertex 1 with 0.

2. i:= 0.

3. Find all unlabelled vertices adjacent to at least one vertex labelled i, and label them with i+1. If there is no such a vertex then return ∞.

4. If vertex n is labelled then return i+1.

5. i:= i+1 and goto 3.

In order to realize the algorithm by an RAM_0 program, we use R_{n+1}, R_{n+2}, ..., R_{n+m} to store the labels on the vertices $1,2,...,m$ respectively. The algorithm can be written as

```
BEGIN
  R[n+1]:= 0;
  FOR i:= 2 upto m DO R[n+i]:= n;
      (label vertex 1 with 0, label other vertices with n)
  FOR i:= 0 upto m DO
    FOR j:= 1 upto m DO
      FOR k:= 1 upto m DO
        IF R[(j-1)m+k]=1 and R[n+j]=i
            (vertices j and k are adjacent and vertex j is labelled i)
        THEN R[n+k]:= Min{i+1, R[n+k]};
            (if vertex k is not labelled then label it with i+1)
  IF R[n+m]=n THEN return ∞ ELSE return R[n+m].
END.
```

If there is a path from 1 to n, the distance must be less than n. $R[(j-1)m+k] = 1$ means that the vertices j and k are adjacent. Then, if vertex k is not labelled, i.e., $R[n+k] = n$, it will be labelled with i+1 by $R[n+k] := Min\{i+1, R[n+k]\}$. Otherwise we must have $R[n+k] \leq i$, the label on vertex k will not be changed.

The i,j,k are all index registers. All elements with subscript can be realized by indirect addressing. The reader can translate the above program to an RAM_0 program directly.

The uniform sequential time complexity is obviously $O(m^3)$. We can improve the algorithm so that the uniform time complexity is only $O(m^2)$. See Exercise 22.3.

EXAMPLE 22.4 Depth First Search

Consider visiting vertices of an undirected graph in the following way. Select and visit a starting vertex v. Then select a neighbour w of v and visit w. In general, suppose that x is the vertex we are visiting.

Then we consider a neighbour of x, say y. If y has been visited then select another neighbour of x. If y has not been visited then we visit y and start the search at y. After completing the search through all paths beginning at y, the search returns to x. Finally the search returns to the starting vertex v, if the graph is finite. Since we continue searching in the deeper direction as long as possible, this method is called a depth first search.

```
PROCEDURE SEARCH (v)
BEGIN
   mark v "old"; print v;
   FOR each neighbour w of v DO
      IF w is new THEN SEARCH (w)
END.
```

To realize this algorithm, we need the following data structure. For each vertex u, there is a list of its neighbours, L_u, which is a segment of an array Neighbour. Arrays First[u] and Last[u] store the first and last addresses of L_u in Neighbour. Array Mark[u] stores the mark (new or old) for u. Furthermore we need two stacks S[i] and T[i] to store the addresses of vertices to be searched in level i. For example, if the graph is as shown by Fig. 22.1 then the arrays are:

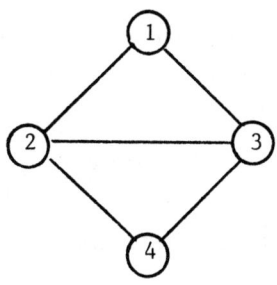

Fig. 22.1

Neighbour [1]	2	First[1]	1	Last[1]	2	Mark[1]	old
[2]	3	[2]	3	[2]	5	[2]	new
[3]	1	[3]	6	[3]	8	[3]	new
[4]	3	[4]	9	[4]	10	[4]	new
[5]	4						
[6]	1						
[7]	2						
[8]	4						
[9]	2						
[10]	3						

The program is thus

```
SEARCH (u)
BEGIN
  i:=1; print u; Mark[u]:=old, S[1]:=First[u]; T[1]:=Last[u];
  WHILE i≠0 DO
    IF S[i]>T[i] THEN i:=i-1 ELSE
      BEGIN s:=S[i]; v:=Neighbour[s];
        IF Mark[v]=old THEN S[i]:=S[i]+1 ELSE
        (print v; Mark[v]:=old; i:=i+1;
          S[i]:=First[v]; T[i]:=Last[v])
      END
END
```

Now the reader can translate the program to an RAM_0 program in a straightforward way. Suppose that the number of vertices and the number of edges in the graph are n and e respectively. Then the uniform space complexity is $O(n+e)$. The uniform time complexity is only $O(e)$.

From the last two examples we can see that the RAM_0 model is widely used in designing algorithms for graphs and matrices, order statistics, set manipulation and many other problems.

Exercises

22.1 A heap is a labelled binary tree in which every leaf is of depth d or d-1. The label on any node is greater than or equal to the labels on its sons. A heap can be represented by an array A. For example, the heap in Fig. 22.2 can be represented by the following array A:

|16|11|9|10|5|6|8|1|2|4|

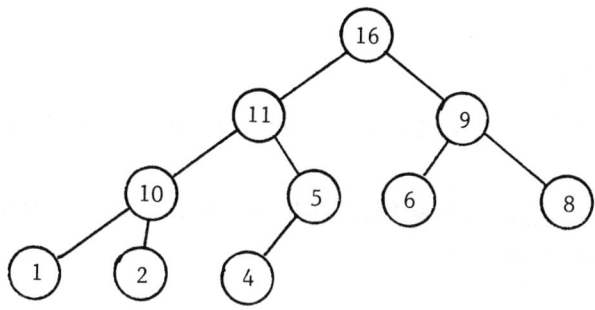

Fig. 22.2

A[2i] and A[2i+1] are the labels on the left-hand and right-hand sons of the node with label A[i].

If the binary tree is arbitrarily labelled, then we can construct a heap by the following algorithm:

PROCEDURE HEAPIFY (i)

    IF i is not a leaf AND IF a son of i has a label greater than the label on i

    THEN BEGIN

        let k be the son of i with the largest label;

        interchange A[i] and A[k];

        HEAPIFY (k)

    END.

FOR i:= n down to 1 DO HEAPIFY (i).

Write a RAM_o for this algorithm. What is the uniform time complexity?

22.2 Use two arrays Leftson and Rightson to represent a binary tree. On the binary tree do the inorder traversal, which is defined recursively:

1. Visit in inorder the left subtree of the root r;
2. Visit r.
3. Visit in inorder the right subtree of the root r.

Design an RAM_0 to do the work.

22.3 Improve the algorithm in Example 22.3 so that the uniform time complexity is only $O(m^2)$.

§23 RAM SIMULATING TM

In this section and the next section, we consider only the first input form of the RAM.

Let $T = (Q, I, \Sigma, \delta, \sqcup, q_0, F)$ be an (f,g)-TM, where $f \leq g^*$. Without loss of generality, we assume that

* $\Sigma = \{0, 1, \ldots, a\}$ where a is the blank $\sqcup$;

* $Q = \{0, 1, \ldots, b\}$ where $q_0 = 0$, $F = \{1\}$;

* tape 1 is the input tape;

* tape $2, \ldots,$ tape $m-1$ are the work tapes;

* tape m is the output tape;

* whenever the work heads and the output head rewrite, they must move;

* all the used work space (on work tapes) are to the right of the initial positions of worktape heads, or equivalently, every worktape has a leftmost square at which is the initial position of the head.

Now let us design an RAM_0 M with k indices ($k > m$) to simulate T.

For a given input $w = a_1 a_2 \ldots a_n$, the possible position of input head is 0(scanning $\sqcup$), 1(scanning a_1), 2(scanning a_2),...,n(scanning a_n), $n+1$(scanning $\sqcup$). The possible position of each work head is from 0 to $s(n) \leq g(n)$.

If we have the following information at a given moment:

(1) in I_i, the position of the i-th head ($i = 1, 2, \ldots, m$).

(2) in I_{m+1}, the state q of T.

(3) in $R_0, R_1, R_2, \ldots, R_n, R_{n+1}$, the input of T: $a, a_1, a_2, \ldots, a_n, a$. (a represents $\sqcup$).

(4) in R_p where $p = n+j+i\cdot(m-1)$, the contents of the i-th square of the j-th tape ($j = 2,3,\ldots,m$; $i = 0,1,2,\ldots$), that is, the 0-th squares of tape $2,3,\ldots,m$ are stored in

$$R_{n+2}, R_{n+3}, \ldots, R_{n+m};$$

the first squares of tape $2,3,\ldots,m$ are stored in

$$R_{n+m+1}, R_{n+m+2}, \ldots, R_{n+2m-1};$$

the second squares of tape $2,3,\ldots,m$ are stored in

$$R_{n+2m}, R_{n+2m+1}, \ldots, R_{n+3m-2};$$

$$\ldots\ldots\ldots\ldots,$$

then, by indirect addressing, RAM_0 can get all information needed to determine the movement at this moment and can simulate one move of T. In detail, it can

* get the state of T and all the symbols scanned by the head of T by indirect read instructions;

* get the move-mode of T (i.e., the value of δ) at this moment by a series of decisions (in the program, a directed graph);

* simulate the rewriting of the heads of T by indirect write instructions;

* simulate the movements of the heads of T by adding or subtracting 1 in I_1 or m-1 in $I_2,\ldots,I_m$.

Since the reading and writing of the indices in M do not interrupt a phase and T does not read a symbol written in the present phase until the next phase has begun, every phase of T can be simulated by M in one phase.

Of course, when M starts its simulation, conditions (1) to (4) do not hold. We must do some preparations:

* set $I_1 = 1$, $I_2 = n+2,\ldots,I_m = n+m$, $I_{m+1} = 0$;

* set $R_0 = a$, $R_1 = a_1$, $R_2 = a_2,\ldots,R_n = a_n$, $R_{n+1} = a$.

95

This preparation can be completed in one phase and $O(n)$ space, if the first input form is given.

Thus this simulating RAM_0 M can be described as

    BEGIN

        preparation;

        REPEAT

            simulate one step of T

        UNTIL T enters state 1

    END.

THEOREM 23.1 If (f,g) is a flat nice pair and T is an (f,g)-TM, then there exists an $(O(f),O(g))$-RAM_0 M (logarithmic cost) simulating T.

Exercises

23.1 Why do we give the following restraint * to T in the proof of Theorem 23.1?

*: whenever the work heads and the output head rewrite, they must move. Without this restraint, give the proof of Theorem 23.1.

23.2 Prove that for any (f,g)-TM M, there is an (f,g)-TM M' simulating M and satisfying all the restraints listed at the beginning of this section.

§24 LSTM SIMULATING RAM

LEMMA 24.1 For any (f,g)-RAM M_1, there is an $(O(f),O(g))$-RAM M_2 which can simulate M_1 and has two special states q' and q'' such that whenever M_2 begins a new phase, it enters one of the special states q' and q'', otherwise it does not enter q' nor q''.

Proof: M_2 has one more index I than M_1 and uses it to record the largest j in the way that R_j has been used so far. Also M_2 uses R_{2i} and R_{2i+1} to simulate R_i of M_1:

    R_{2i} of M_2 stores the contents of R_i of M_1;

R_{2i+1} of M_2 stores 1 or 0 indicating whether or not the contents of R_{2i} are written in the present phase.

When M_1 writes a number into its i-th register, M_2 writes the same number into its (2·i)-th register and writes number 1 into its (2·i+1)-th register. When M_1 reads a number from its i-th register, M_2 checks if its (2·i+1)-th register is 0. If so, M_2 reads the contents from R_{2i}. Otherwise, a phase of M_1 must be interrupted, M_2 enters its special state q' and fills all the odd registers with 0 (by means of the extra index I), then enters state q" and starts to simulate the next phase of M_1. The space consumption of index I is obviously O(log g) (see Exercise 21.5).

Now we proceed to design an LSTM to simulate a given RAM.

Let M be an RAM with the property described in Lemma 24.1, with reversal r(n) and space s(n). Then for any ID PQ, where P is the prefix and Q is the suffix (see §21), we have that $|P|$ = O(log s(n)) and $|Q|$ = O(s(n)·log s(n)).

The key problem is: given an ID at the time when M starts a new phase, how to find the ID at the beginning of the next phase by means of an LSTM. We divide the solution of this problem into several steps.

Because we do not know the space complexity s(n) of M, we choose an integer s arbitrarily and suppose that s is s(n). Whenever we find s is too small, we will try a larger one.

(1) find-all-prefixes

INPUT: $PQ,1^s$ (where "," is a symbol in the input string, as a separator.)

(24.1)

OUTPUT: $P_1Q,P_2Q,\ldots,P_tQ,PQ,1^s$ (24.2)

here t = $O(s^*)$ and $P_1,P_2,P_3,\ldots,P_t$ are all possible prefixes of length not exceeding c·(log s).

This can be realized by an LSTM of phase 1 and width $O(s^*(n))$. In fact, what is needed is to output all strings of length not exceeding c·(log s) in lexicographic order, copying Q after each string.

(2) calculate-next-prefixes

INPUT: (24.2)

OUTPUT: $(P_1,P_1',n_1,c_1),(P_2,P_2',n_2,c_2),\ldots,(P_t,P_t',n_t,c_t),PQ,1^s$ (24.3)

where (P_i,P_i',n_i,c_i) indicates that if the ID of M is P_iQ, then, in one step, the prefix of the ID will become P_i' and the contents of the n_i-th ordinary register of M will be changed to c_i.

Obviously, since any instruction of the RAM is in NC there is a $(\log^* m, m^*)$-LSTM which transforms any ID P_iQ of M to (P_i,P_i',n_i,c_i) where $m = |P_iQ|$ is the input length and P_i', n_i, c_i have the same meaning as mentioned above. According to Theorem 18.1, we know the desired LSTM is of phase $\log^*(s(n))$ and width $s^*(n)$.

(3) find-phase-path

INPUT: (24.3)

OUTPUT: $(P_1'',n_1',c_1'),(P_2'',n_2',c_2'),\ldots,(P_p'',n_p',c_p'),Q,1^s$ (24.4)

or Error

where Error indicates that s is too small, and in (24.4)

P_1'' is P;

P_{i+1}'' is the next prefix of P_i'';

P_p'' is the prefix when the phase ends;

the "changing information" (n_i', c_i') has the same meaning as in 2) (see Fig. 24.1).

$$R_{n_1'} := c_1' \qquad R_{n_2'} := c_2' \qquad R_{n_p'} := c_p'$$

$$P = P_1'' \dashrightarrow P_2'' \dashrightarrow P_3'' \ldots P_{p-1}'' \dashrightarrow P_p''$$

Fig. 24.1 A phase-path

Without loss of generality, we may assume that $P = P_1$ in (24.3). If s is large enough, then

P_1', the next prefix of $P_1 = P$, must be some P_i

P_i', the next prefix of P_i, must be some P_j

P_j', the next prefix of P_j, must be some P_k

....

Thus we can obtain a chain of prefixes or a phase-path:

$$P \to P_i \to P_j \to P_k \to \ldots\ldots,$$

step by step, until the present phase ends (because of the property mentioned in Lemma 24.1, it is easy to determine when a phase ends). Since, in one phase, the RAM does not read what is written within the same phase, the information (n_i, c_i) obtained from the unchanged suffix Q, is correct until the present phase ends. By means of the information in (24.3), an LSTM can get (24.4) in phase 1 and width $s^*(n)$.

But if s chosen in advance is not large enough, then, while finding the phase path, there may be some moment at which we cannot find the next prefix in (24.3). Then the LSTM outputs Error.

(4) substitute

INPUT: (24.4)

OUTPUT: $P_1 Q_1, 1^s$ (24.5)

where P_1 is P_p'' in (24.4) and Q_1 is the result of substituting all the "changing information" in (24.4) into Q.

The contents of an ordinary register may be changed several times. When the phase ends, the contents of this register should be the contents after the last change. Once a symbol is written on the output tape, it cannot be changed. So the LSTM should work in the following way.

For each ordinary register R_j used, the LSTM looks for the name j in (24.4) from the right to the left. Once this name j appears in (24.4) (excluding Q), the LSTM writes the corresponding new contents after j on the output tape immediately. If the name does not appear in (24.4) (excluding

Q), then the LSTM copies the corresponding contents in Q in (24.4) onto the output tape.

This can be done by an LSTM of phase 1 and width $s^*(n)$.

Summarizing the above four steps, we know that if s is large enough, an LSTM can transform $PQ,1^S$ to $P_1Q_1,1^S$ within phase $\log^*s(n)$ and width $s^*(n)$, where PQ is the ID of the RAM at the beginning of a phase and P_1Q_1 is the ID of the RAM at the end of this phase.

In order to simulate the RAM, the LSTM must

(*) initiate,

that is, transform the input of the RAM into $P_0Q_0,1^S$ where P_0Q_0 is the initial ID of the RAM.

Of course, the LSTM must also be able to check the conditions

(*) halt, and

(*) error.

Finally, the LSTM must have the ability to

(*) transform-output,

that is, when the LSTM gets $PQ,1^S$ and PQ is a halting ID of the RAM, it picks up the contents of the output registers of the RAM and writes these contents on its output tape to form its own output.

Now the simulating LSTM can be expressed as

BEGIN

(1)   s:=1;

(2)   REPEAT

(2.1)   s:=2*s;

(2.2)   initiate;

(2.3)   REPEAT

(2.3.1)   find-all-prefixes;

(2.3.2)   calculate-next-prefixes;

(2.3.3)   find-phase-path;

(2.3.4)   IF not-Error THEN substitute

          UNTIL halt or Error

       UNTIL halt;

(3)   transform-output

       END.

Notice (2.1) to (2.3) will be executed at most log s(n) times. For a given s, (2.3.1) to (2.3.4) will be executed at most r(n) times. So the phase of this LSTM is $r(n) \cdot \log^* s(n)$, the width is $s^*(n)$.

THEOREM 24.1   If M is an RAM of reversal r(n) and space s(n), then there is an LSTM of phase $r(n) \cdot \log^* s(n)$ and width $s^*(n)$ simulating M.

If M is an (f,g)-RAM with $f \leq g^*$ (flat), then the phase of the LSTM is

$$f \cdot \log^* g = O(f^*),$$

and the width is g^*. Thus we have

THEOREM 24.2   For any (f,g)-RAM with $f \leq g^*$, there is an (f^*, g^*)-LSTM simulating it.

Up till now, in the flat case we have proved that

$$TM \leq {}^*RAM_0 \leq {}^*RAM \leq {}^*LSTM$$

Exercises

24.1   Prove that for an (f,g)-RAM M, the strategy of "try different s" is not necessary in the design of the simulating LSTM.

24.2   Prove that $t(n) = O(2^{s^*(n)})$ for any RAM.

24.3   Design an RAM such that

$$r(n) = n, \quad s(n) = 2^{2^n}.$$

Notice that this is not a nice pair and we will not consider this kind of RAM.

6 Vector machines

§25 DEFINITION

The vector machine proposed by Pratt and Stockmeyer is a famous and important computational model. As a parallel model, it is quite different from sequential models such as TM or RAM. We now introduce it.

I. Vectors

If $a_i \in \{0,1\}$, $i = 0,1,2,\ldots$, and there is a natural number m such that $a_i = a_{i+1}$ ($i = m, m+1,\ldots$), then we call the sequence (infinite to the left)

$$\ldots a_m a_{m-1} \ldots a_2 a_1 a_0$$

a vector, and call a_i the i-th term of the vector.

In other words, a vector is an ultimately constant $\{0,1\}$ sequence. Exept for a finite number of terms, its terms are all equal.

For example, $\ldots 0010010$ is a vector, whose first and fourth terms are 1, and all other terms are 0.

Suppose that $V = \ldots a_3 a_2 a_1 a_0$ is a vector. Let n be the minimum integer m mentioned in the definition, i.e.

$$n = \text{Min } \{m | a_m = a_{m+1} = a_{m+2} = \ldots\}.$$

We define $|V| = \text{length}(V) = n$, and

$$\text{number}(V) = \begin{cases} + \sum_{i=0}^{n-1} a_i \cdot 2^i & \text{when } 0 = a_n = a_{n+1} = \ldots \\ \\ - \sum_{i=0}^{n-1} a_i \cdot 2^i & \text{when } 1 = a_n = a_{n+1} = \ldots \end{cases}$$

Thus, every vector represents an integer. For example,

number(...001001) = + 9,   length(...001001) = 4

number(...11100101) = -5,   length(...11100101) = 5

number(...1110101) = -5,   length(...1110101) = 4

number(...1...111) = 0,   length(...1...111) = 0

Suppose that $A = ...a_2a_1a_0$, $B = ...b_2b_1b_0$ are two vectors; we define their Boolean operations as follows

$A \wedge B = ...c_2c_1c_0$, where $c_i = a_i \wedge b_i$, $i = 0,1,2,...$,

$\neg A = ...d_2d_1d_0$,   where $d_i = \neg a_i$, $i = 0,1,2,...$,

i.e., the operations of two vectors are defined by Boolean operations on their corresponding bits. The other Boolean operations can be defined by means of $\neg$ and $\wedge$:

$A \vee B = \neg(\neg A \wedge \neg B)$,

$A \oplus B = (\neg A \wedge B) \vee (A \wedge \neg B)$,

$A \leftrightarrow B = (\neg A \vee B) \wedge (A \vee \neg B)$,

and so on.

In the following, we use + to represent a sequence ...000, - to represent a sequence ...111. Thus a vector can be represented by +w or -w, $w \in \{0,1\}^*$. Obviously we have

+w = + 0w,

-w = - 1w.

II. Constituents of a vector machine

(1) k registers $V_1, V_2, ..., V_k$, where k is a constant number. Each register can store a vector (the contents of the register). We often do not distinguish between the register and its contents. For example, by vector V_1 we mean the contents of register V_1, by "the vector changes" we mean the contents of the register changes.

103

(2) A program, which is a directed graph (the same as that of RAM). The nodes of the graph are the states. Each node has a fan-out of at most 2. If the fan-out of a node is 0, then the node is a halting state. If the fan-out is two, then two opposite predicates $V_i = +$ and $V_i \neq +$ should be labelled on the two edges out from the node. If the fan-out is 1, then an instruction should be labelled on the edge out from the node. A designated node q_0 is called the initial state.

An instruction has one of the following four forms:

(1) A := a, meaning that the contents of register A should become to the vector a;
(2) A := B ∧ C;
(3) A := ¬B;
(4) A := B ↑ C (or A := B ↓ C).

Instruction (4) means that the contents in register B should be shifted to the left a distance of number (C), and put into A. The register B does not change. (If number (C) < 0, then it should be shifted to the right a distance of -number(C).) When shifting left, 0's will be added to the right-hand end automatically. Instruction A := B ↓ C means shifting a distance of number (C) to the right.

For example, $V_1 := + 10$ means to put +10 into V_1. When $V_2 = -00110$, $V_0 = +11$, $V_1 := V_2 \uparrow V_0$ means to put -00110000 into V_1. When $V_2 = -00110$, $V_0 = -0011$, $V_1 := V_2 \uparrow V_0$ means to put -00 into V_1.

III. Execution of the program

The input of the VM is a vector stored initially in V_1. When the machine starts, the contents of other registers $V_2, V_3, \ldots, V_k$ are all +'s (+ = ...000), and the machine is in the initial state q_0.

Generally, the machine is in a state q. If the fan-out of q is 1 then the machine executes the instruction labelled on the edge out from q (therefore changes the contents in some register), and goes into the next state in the graph. If the fan-out number of q is 2, then the machine goes into the next state along the edge on which the predicate is satisfied. If the fan-out number of q is 0, then the machine halts and the contents of some registers will be considered as the output.

IV. The space

The input word w of VM is a binary string of length n, i.e., $w \in \{0,1\}^n$. If $w \in 0 \cdot \{0,1\}^{n-1}$ then $-w$ is initially in V_1. If $w \in 1 \cdot \{0,1\}^{n-1}$, then $+w$ is initially in V_1. Therefore initially we have $|V_1| = n$.

For input w, the sum over all $i = 1,2,\ldots,k$ of the maximum lengths of the contents stored in V_i, during the whole computation, is denoted by $S(w)$. The function $S(n) = \text{Max}\{S(w) \mid |w| \leq n\}$ is called the space complexity of the vector machine.

V. The (parallel) time

For input w, the total number of times that the VM executes an instruction or tests a predicate is denoted by $T(w)$. The function $T(n) = \text{Max}\{T(w) \mid |w| \leq n\}$ is called the (parallel) time complexity of the VM.

We only consider the VM's with $T(n) < \infty$, i.e., the VM that halts for any input. In this case it is easy to see $S(n) < \infty$.

Exercises

25.1 Prove that if $x = \text{number}(A) > 0$ then $|A|$ is exactly the length of the binary expression of x. Is this still true when $x \leq 0$?

25.2 Suppose that $V_1 = +11110000$, $V_2 = -010$. After executing the following instructions,

$V_0 := V_1 \uparrow V_2$

$V_1 := V_1 \oplus V_0$,

what are the contents in V_1?

25.3 Suppose that A and B are two vectors, number$(A) = a$ and number$(B) = b$.
 (1) If $a = b > 0$ then $A = B$.
 (2) If $a = b \leq 0$, is $A = B$ always true?

25.4 If we define the time complexity to be the total number of times that the machine executes instructions, or the total number of times that the machine tests the predicates, what is the relation between the definition and the original definition?

§26 EXAMPLES

EXAMPLE 26.1 In the previous section we have mentioned the instruction A := B ↑ C which means to shift the contents of B left a distance given by number(C). When number(C) > 0, the vacated positions are filled with 0's. We execute the following six instructions. The result is to shift B left a distance given by number(C), but the vacated positions are filled with 1's. We shall denote the sequence of instructions

$$X := -;$$
$$X := X \leftarrow C;$$
$$X := \neg X;$$
$$A := B \uparrow C$$
$$A := A \vee X$$
$$X := +$$

by A := B ↑↑ C. Sometimes we write an instruction such as A := B ↑↑ 5. The meaning is clear.

EXAMPLE 26.2 Function $\lceil \log(n+1) \rceil$,

INPUT: $V_1 = + 1^n$, the unary representation $U(n)$ of n.

OUTPUT: $V_2 = 1^{\lceil \log(n+1) \rceil}$, $U(\lceil \log(n+1) \rceil)$.

COMPLEXITY: time = $O(\log n)$, space = $O(n)$.

We product 1^{n+1} in $V_1(V_1 := V_1 \uparrow\uparrow 1)$. Then try to shift V_1 to the right a distance $1, 2, 4, \ldots, 2^m$ until we obtain a +, then $m = \lceil \log(n+1) \rceil$ (because $2^{m-1} < n+1 \leq 2^m$). The program is then

BEGIN

$V_1 := V_1 \uparrow\uparrow 1;$

$V_2 := -01;$

WHILE $V_1 \uparrow V_2 \neq +$ DO $V_2 := V_2 \uparrow 1;$

(now $V_2 = -010^m$)

$V_2 := V_2 \vee (V_2 \uparrow 1)$ $(V_2 = -0^m)$

$V_2 := \neg V_2$ $(V_2 = +1^m)$

END.

EXAMPLE 26.3 Transform unary representation to binary representation.

INPUT: $V_1 = +1^n$, the unary representation $U(n)$.

OUTPUT: $V_2 = + a_m a_{m-1} \ldots a_2 a_1 a_0$ (when $m \neq 0$, $a_m \neq 0$, and $\Sigma a_i 2^i = n$), the binary representation $B(n)$.

COMPLEXITY: time = $O(\log n)$, space = $O(n)$.

When $n \neq 0$, the length of the binary representation of n is $m+1 = \lceil \log(n+1) \rceil$. By Example 26.2, we can obtain $+1^{\lceil \log(n+1) \rceil}$. Then determine whether a_i is 0 or 1 for $i = m, m-1, \ldots, 2, 1, 0$. a_m is always 1. To determine a_{m-1}, we first obtain 1^{n-2^m}. Then check if $n-2^m < 2^{m-1}$. If it is, then $a_{m-1} = 0$ else $a_{m-1} = 1$. This way, we can obtain the binary representation within time $O(\log n)$ and space $O(n)$. The program is

BEGIN

$Y := +1 \ 0^{\lceil \log(n+1) \rceil}$; $V_2 := +$; $V_1 := V_1 \uparrow\uparrow 1$;

WHILE $Y \neq +$ DO (if $V_1 \downarrow Y \neq +$ THEN ($V_2 := V_2 \uparrow\uparrow 1$; $V_1 := V_1 \downarrow Y$) ELSE

$V_2 := V_2 \uparrow 1$; $Y := Y \downarrow 1$)

END.

EXAMPLE 26.4 Copping (V_1, m, ℓ)

INPUT: $V_1 = w$

$V_2 = 1^{|w|} = 1^\ell$

$V_3 = 1^m$

OUTPUT $V_4 = (w)^m$

COMPLEXITY: time = $O(\log \ell + \log m)$, space = $O(m\ell)$

We compute the binary representation of m and ℓ, and store in V_5 and V_6 respectively. Using V_6 we can double the contents in V_7. Then determine whether or not we should concatenate the contents of V_7 to the right-hand side of V_4, according to whether the corresponding bit of V_5 is 1 or 0. The program is

    BEGIN

        $V_5 := B(m); V_6 := B(\ell); V_7 := V_1; V_8 := +1;$

        WHILE $V_5 \neq +$ DO

        BEGIN

          IF $V_5 \wedge V_8 \neq +$ THEN $(V_4 := V_4 \uparrow V_6; V_4 := V_4 \vee V_7)$

          $V_7 := V_7 \vee (V_7 \uparrow V_6)$ (double V_7)

          $V_6 := V_6 \uparrow 1$ (the length of V_7)

          $V_5 := V_5 \downarrow 1$

        END

    END.

The first and second statements use time $O(\log m)$ and time $O(\log \ell)$ respectively. The while statement will be executed $\log m$ times, each time $O(1)$ instructions. Therefore the total time is $O(\log m + \log \ell)$. The space is obviously $O(m\ell)$.

It is easy to see that if

$$V_1 = 0^{(m-1)\ell} w_1 0^{(m-1)\ell} w_2 \ldots 0^{(m-1)\ell} w_n$$

$(|w_i| = \ell, i = 1, 2, \ldots, n)$

then the output will be

$$V_4 = w_1^m w_2^m \ldots w_n^m.$$

EXAMPLE 26.5 Copping' (V_1, m, ℓ).

    INPUT: $V_1 = w$

              $V_2 = 1^{|w|} = 1^\ell$

              $V_0 = 1^m$

OUTPUT: $V = w^{2^m}$

COMPLEXITY: time = $O((\log \ell) + m)$, space = $O(\ell \cdot 2^m)$

The program is simply Copping $(V_1, 2^m, \ell)$.

EXAMPLE 26.6 Reversed word

INPUT: $V_1 = a_1 a_2 \ldots a_n$

$V_2 = 1^n$, where $n = 2^\ell$.

OUTPUT: $a_n \ldots a_2 a_1$

COMPLEXITY: time = $O(\log n)$, space = $O(n)$.

We define graduated rulers $\mu_{\ell j}$ $(0 \leq j \leq \ell-1)$ as $\mu_{\ell j} = (1^{2^j} 0^{2^j})^{2^{\ell-j-1}}$.

For example

μ_{40} = 1010101010101010

μ_{41} = 1100110011001100

μ_{42} = 1111000011110000

μ_{43} = 1111111100000000.

Given $X = +1^\ell$, we can construct $\mu_{\ell,\ell-1}$ in $O(1)$ time and $O(2^\ell)$ space:

Y := +

Y := Y ↑↑ $2^{\ell-1}$

Y := Y ↑ $2^{\ell-1}$

The integer $2^{\ell-1}$ (which is $+10^{\ell-1}$) can be obtained from X in constant time. In general, given $X = \mu_{\ell,j}$ we can construct $\mu_{\ell,j-1}$ as follows

Z := X ↑ (-2^{j-1})

X := X ⊕ Z.

Now we compute the reversed word. Consider the case when $n = 8$. We do 4-4 exchange to obtain $a_5 a_6 a_7 a_8 a_1 a_2 a_3 a_4$. This can be done with $\mu_{3,2}$ in $O(1)$ time:

$$V_1 := (V_1 \wedge \mu_{3,2}) \downarrow 2^2 \vee (V_1 \wedge \neg\mu_{3,2}) \uparrow 2^2.$$

Then do 2-2 exchange to get $a_7 a_8 a_5 a_6 a_3 a_4 a_1 a_2$ by $\mu_{3,1}$:

$$V_1 := (V_1 \wedge \mu_{3,1}) \downarrow 2^1 \vee (V_1 \wedge \neg\mu_{3,1}) \uparrow 2^1.$$

Finally do 1-1 exchange to obtain the inversed word $a_8 a_7 a_6 a_5 a_4 a_3 a_2 a_1$ by $\mu_{3,0}$:

$$V_1 := (V_1 \wedge \mu_{3,0}) \downarrow 2^0 \vee (V_1 \wedge \neg\mu_{3,0}) \uparrow 2^0.$$

Generally when $n = 2^\ell$, the program is

$$X := \mu_{\ell,\ell-1}; \quad Y := + 10^{\ell-1};$$

$$\text{WHILE } Y \neq + \text{ DO } (V_1 := (V_1 \wedge X) \downarrow Y \vee (V_1 \wedge \neg X) \uparrow Y; \quad Y := Y \downarrow 1;$$

$$X := X \oplus (X \downarrow Y)).$$

To construct $\mu_{\ell,\ell-1}$, the time used is $O(\log n)$. The WHILE statement will be executed $O(\log n)$ times. Therefore we have the complexities desired.

If n is not a power of 2, we can add some 0's to the right-hand side so that the length becomes a power of 2. (Exercise: How is this done?). Then compute the inversed word.

EXAMPLE 26.7 All codings of length n

  INPUT $V_1 = 1^n$.

  OUTPUT V_2 = all binary sequences of length n.

  COMPLEXITY   time = $O(n)$, space = $O(n \cdot 2^n)$.

For example, if n = 3 and we write the codings in column, then all the codings of length 3 are

    11110000

    11001100

    10101010.

The three rows are exactly $\mu_{3,2}$, $\mu_{3,1}$, $\mu_{3,0}$. We generate graduated rulers with "holes" $K_{n,j}$ ($0 \leq j \leq n-1$)

$$K_{n,j} = ((1 \cdot 0^{n-1})^{2^j} (0^n)^{2^j})^{2^{n-j-1}}.$$

For example

$$K_{3,2} = 100\ 100\ 100\ 100\ 000\ 000\ 000\ 000$$
$$K_{3,1} = 100\ 100\ 000\ 000\ 100\ 100\ 000\ 000$$
$$K_{3,0} = 100\ 000\ 100\ 000\ 100\ 000\ 100\ 000$$

Then the output should be $k_{3,2} \vee k_{3,1} \downarrow 1 \vee k_{3,0} \downarrow 2$. Thus we have the following program:

BEGIN

    $V_2 := K_{n,n-1}$

    FOR $j := 1$ UNTIL $n-1$ DO $V_2 := K_{n,n-1-j} \downarrow j \vee V_2$

END.

To construct $K_{n,n-1}$ from $V_1 = 1^n$ needs time $O(\log n)$. To construct $K_{n,j-1}$ from $K_{n,j}$ needs time $O(1)$.

EXAMPLE 26.8 Holing (V_1,n,r,s)

    INPUT $V_1 = \alpha_1 \alpha_2 \ldots \alpha_n$, $(|\alpha_i| = r)$

             $V_2 = 1^n$

             $V_3 = 1^r$

             $V_4 = 1^s$

    OUTPUT $\alpha_1 0^s \alpha_2 0^s \ldots \alpha_n 0^s$.

    COMPLEXITY: time $= O(\log(nsr))$, space $= O((r+s)n)$.

For simplicity, we abbreviate the subscripts and assume that n is a power of 2.

We can obtain $B(n)$, $B(s)$, $B(r)$, $B(sn)$, $B(rn)$ within time $O(\log(nsr))$. With them we can construct

(1) in constant time

$$\overbrace{\alpha\ldots\alpha}^{n}0^s0^s\overbrace{0\ldots0}^{n}{}^s$$

and ruler

$$\overbrace{0^r\ldots\ldots0^r}^{n/2}\overbrace{1^r\ldots\ldots1^r}^{n/2}0^s0^s\overbrace{\ldots\ldots\ldots0^s}^{n},$$

(2) in another constant time

$$\overbrace{\alpha\ldots\ldots\alpha0^s}^{n/2}\overbrace{\ldots\ldots\ldots0^s}^{n/2}\overbrace{\alpha\ldots\ldots\alpha0^s}^{n/2}\overbrace{\ldots\ldots0^s}^{n/2}$$

and ruler

$$\overbrace{0^r\ldots\ldots0^r}^{n/4}\overbrace{1^r\ldots\ldots1^r}^{n/4}\overbrace{0^s\ldots\ldots0^s}^{n/2}\overbrace{0^r\ldots\ldots0^r}^{n/4}\overbrace{1^r\ldots\ldots1^r}^{n/4}\overbrace{0^s\ldots\ldots\ldots0^s}^{n/2},$$

(3) ..., and so on.

The program is

   BEGIN

      I := B(n/2); J := B(nr/2); K := B(ns); X := +; X := (X↑↑J)↑K; V_1 := V_1↑K;

      WHILE I ≠ + DO (J := J ↓ 1; K := K ↓ 1; I := I ↓ 1;

                 V_1 := (V_1 ∧ X) ↓ K ∨ (V_1 ∧ ¬X);

                 Y := X ⊕ (X ↑ J);

                 X := (Y ∧ X) ↓ K ∨ (Y ∧ ¬X);

   END

(X is the ruler).

EXAMPLE 26.9 Alternative connection.

   INPUT $V_1 = \alpha_1\alpha_2\ldots\alpha_n$ ($|\alpha_i| = r$)

             $V_2 = \beta_1\beta_2\ldots\beta_n$ ($|\beta_i| = s$)

             $V_3 = 1^r$

$$V_4 = 1^s$$
$$V_5 = 1^n$$

OUTPUT $V_6 = \alpha_1\beta_1\alpha_2\beta_2\ldots\alpha_n\beta_n$.

COMPLEXITY: time = $O(\log(nsr))$, space = $O((r+s)n)$.

We use $O(\log(nsr))$ time to construct

$$\alpha_1 0^s \alpha_2 0^s \ldots \alpha_n 0^s$$

$$0^r\beta_1 0^r\beta_2 \ldots 0^r\beta_n.$$

The program is simply

$$V_2 := \text{Holing}(V_1,n,r,s) \vee \text{holing}(V_2,n,s,r) \downarrow r$$

EXAMPLE 26.10 Direct product connection.

INPUT $V_1 = \alpha_1\alpha_2\ldots\alpha_n$, $V_2 = \beta_1\beta_2\ldots\beta_m$, $(|\alpha_i| = r, |\beta_i| = s)$ and $1^r, 1^s, 1^n, 1^m$.

OUTPUT $\alpha_1\beta_1\alpha_1\beta_2\ldots\alpha_1\beta_m\alpha_2\beta_1\alpha_2\beta_2\ldots\alpha_2\beta_m\ldots\alpha_n\beta_1\alpha_n\beta_2\ldots\alpha_n\beta_m$

COMPLEXITY: time = $O(\log(nmsr))$, space = $O((s+r)mn)$.

(1) We use $O(\log(nmsr))$ time to obtain

$$\alpha_1 0^{ms+(m-1)r} \alpha_2 0^{ms+(m-1)r} \ldots \alpha_n 0^{ms+(m-1)r},$$

(2) Use $O(\log(msr))$ time to obtain

$$\alpha_1 0^s \alpha_1 0^s \ldots \alpha_1 0^s \alpha_2 0^s \alpha_2 0^s \ldots \alpha_2 0^s \ldots \alpha_n 0^s \alpha_n 0^s \ldots \alpha_n 0^s,$$

(3) Use $O(\log(msr))$ time to get

$$0^r\beta_1 0^r\beta_2 \ldots 0^r\beta_m$$

(4) Use $O(\log(mnsr))$ time to construct

$$0^r\beta_1 0^r\beta_2 \ldots 0^r\beta_m 0^r\beta_1 0^r\beta_2 \ldots 0^r\beta_m \ldots 0^r\beta_1 0^r\beta_2 \ldots 0^r\beta_m.$$

(5) Putting the results in steps 2 and 4 together, we obtain

$$\alpha_1\beta_1\alpha_1\beta_2\ldots\alpha_1\beta_m\alpha_2\beta_1\alpha_2\beta_2\ldots\alpha_2\beta_m\ldots\alpha_n\beta_1\alpha_n\beta_2\ldots\alpha_n\beta_m.$$

The whole program is thus

$V_1 := \text{Holing}(V_1,n,r,ms+(m-1)r); V_1 := V_1 \downarrow (m-1)(r+s);$

$V_1 := \text{Copping}(V_1,m,r+s);$

$V_2 := \text{Holing}(V_2,m,s,r); V_2 := V_2 \downarrow r;$

$V_2 := \text{Copping}(V_2,n,(r+s)m)$

$V_3 := V_2 \vee V_1.$

EXAMPLE 26.11 Filling.

INPUT $V_1 = w_1w_2\ldots w_m$ ($|w_i| = k$, $i = 1,2,\ldots,m$)

$V_2 = 1^m$

$V_3 = 1^k$

OUTPUT $V_1 = v_1v_2\ldots v_m$, where $v_i = \begin{cases} 0^k & \text{when } w_i = 0^k \\ 1^k & \text{when } w_i \neq 0^k \end{cases}$.

COMPLEXITY: time = $O(\log(mk))$, space = $O(mk)$.

The program is

PROCEDURE Fill_1 (V_1,m,k)

BEGIN

$V_1 := \text{Fill}_1\text{-left}(V_1,m,k)$

$V_1 := \text{Fill}_1\text{-right}(V_1,m,k)$

END.

Where the sub-program Fill_1-left is to extend the 1's to the left in all intervals of length k, while Fill_1-right is to extend the 1's to the right.

PROCEDURE Fill_1-left (V_1,m,k)

BEGIN

$V := (1^{k-1}0)^m$

$I := +1$

DO $\lceil \log k \rceil$ times

$(V_1 := V_1 \vee ((V_1 \uparrow I) \wedge V));$

$V := V \wedge (V \uparrow I);$

$I := I \uparrow 1)$

END.

In the same way we can write a program $Fill_0(V_1,m,k)$ which extends the 0's to every interval of length k.

Exercises

26.1 Design VM:

(1) INPUT $1^n 0 1^m$

OUTPUT 1^n and 1^m

COMPLEXITY: time = $O(\log(m+n))$, space = $O(n+m)$

(2) INPUT $a_1 a_2 \ldots a_{2n-1} a_{2n}$ and 1^{2n}

OUTPUT $a_1 a_3 a_5 \ldots a_{2n-1}$

COMPLEXITY: time = $O(\log n)$, space = $O(n \log n)$.

(3) Input and output are the same as in (2),

COMPLEXITY: time = $O(\log^2 n)$, space = $O(n)$.

(4)* Can you design a VM with time = $O(\log n \cdot \log\log n)$ and space = $O(n \cdot \log\log n)$ for the same problem?

26.2 Design VM:

(1) to test whether or not the sign of a vector is positive.

COMPLEXITY: time = $O(\log n)$, space = $O(n)$,

(2) for the same problem, time = $O(1)$, how much space do you need?

26.3 Prove that the instruction $A := B \downarrow C$ can be realized by other instructions in time $O(\log \text{length}(C))$. When the unary representation $U(\text{length}(C))$ or binary representation $B(\text{length}(C))$ is given, the instruction $A := B \downarrow C$ can be realized in $O(1)$ time by other instructions.

26.4 Design VM:

(1) to test $|V_1| \leq |V_2|$ or $|V_1| > |V_2|$ in time $O(\log(|V_1| \cdot |V_2|))$ and space $O(|V_1| + |V_2|)$.

(2) to test $\text{number}(V_1) \leq \text{number}(V_2)$ in $O(1)$ time.

26.5 Design a VM:

INPUT 1^n

OUTPUT $(10^n)^{n-1} 1$

COMPLEXITY: time = $O(\log n)$, space = $O(n^2)$

26.6* Design a VM:

INPUT 1^n and $a_1 a_2 \ldots a_n$

OUTPUT $a_1 0 a_1 a_2 0 a_1 a_2 a_3 0 \ldots 0 a_1 a_2 a_3 \ldots a_n$

COMPLEXITY: time = $O(\log^* n)$, space = $O(n^2)$.

26.7* Design a VM to add two integers of length n in binary form within time $O(\log n)$ and space $O(n)$.
[Hint: suppose the two integers are in A and B.

$C := B \vee A$

$D := B \wedge A$

$E := +1$

DO log n times {$D := (C \wedge (D \uparrow E)) \vee D$;

$\qquad\qquad\qquad C := (C \uparrow E) \wedge C$;

$\qquad\qquad\qquad E := E \uparrow 1$}

return $A \oplus B \oplus (D \uparrow 1)$.]

26.8* Design a VM to multiply two integers of length n in binary form, within time $O(\log n)$ and space $O(n^2)$. [Hint: add n numbers together by the following "3 to 2" addition method:
Set $S = X \oplus Y \oplus Z$, $C = ((X \wedge Y) \vee (Y \wedge Z) \vee (Z \wedge X)) \uparrow 1$,
Then $X + Y + Z = S + C$.]

26.9 Design a VM:

(1) INPUT $V = \ldots 10^m$

OUTPUT 1^m

COMPLEXITY: time = $O(\log|V|)$, space = $O(|V|)$.

(2) INPUT $V \in +1\{0,1\}^{n-1}$

OUTPUT $+1^n$

COMPLEXITY: time = $O(\log n)$, space = $O(n)$.

26.10 Design a VM,

INPUT $\alpha_1 0^s \alpha_2 0^s \ldots \alpha_n 0^s$ ($|\alpha_i| = r$)

and $1^r, 1^s, 1^n$

OUTPUT $\alpha_1 \alpha_2 \ldots \alpha_n$

COMPLEXITY: time = $O(\log^2(nsr))$, space = $O((r+s)n)$.

26.11 Design a VM Fill(X,m,k,e):

INPUT $X = w_1 w_2 \ldots w_m$ ($|w_i| = ke$) and $1^m, 1^k, 1^e$.

There are m intervals of length ke. Each interval has k segments of length e. In each interval there is at most one segment which is not 0^e.

OUTPUT extend the non-zero segment to the whole interval. (For example, if

$x = 0'0'\ldots a_i'0'\ldots0'0'0'\ldots a_2'\ldots0'\ldots0'0'\ldots a_m'\ldots0'$, $|a_i'| = e, 0' = 0^e$,
$\overbrace{}^{k}\overbrace{}^{k}\overbrace{}^{k}$

then the output is

$a_1 a_1' \ldots a_1' a_2' a_2' \ldots a_2' \ldots a_m' a_m' \ldots a_m'$.)
$\overbrace{}^{k}\overbrace{}^{k}\overbrace{}^{k}$

COMPLEXITY: time = $O(\log(mke))$, space = $O(mke)$.

§27 MATRIX TRANSPOSE AND WORD PROJECTION

In this section two important algorithms for VM are introduced: the matrix transpose algorithm and the word projection algorithm.

Let

$$A = \begin{pmatrix} a_{11} & \cdots & a_{1m} \\ a_{m1} & \cdots & a_{mm} \end{pmatrix}$$

be an $m \times m$ Boolean matrix, ($a_{ij} = 0,1$). In vector machines, we often store A in rows, i.e., the elements of matrix A are arranged and stored in a register in the following way:

$$a_{11}a_{12}\cdots a_{1m}a_{21}a_{22}\cdots a_{2m}\cdots a_{m1}a_{m2}\cdots a_{mm}.$$

The vector $+1^m$ is stored in another register. Now we want to get the transpose matrix A^t, i.e., to output the following vector

$$a_{11}a_{21}\cdots a_{m1}a_{12}a_{22}\cdots a_{m2}\cdots a_{1m}a_{2m}\cdots a_{mm}.$$

We only consider the case when $M = 2^\ell$. Assume that

$$A = \begin{pmatrix} X & Y \\ Z & U \end{pmatrix}, \text{ where } X, Y, Z, U \text{ are } 2^{\ell-1} \times 2^{\ell-1} \text{ matrices.}$$

Then

$$A^t = \begin{pmatrix} X^t & Z^t \\ Y^t & U^t \end{pmatrix}.$$

Therefore we may compute $\begin{pmatrix} X & Z \\ Y & U \end{pmatrix}$ from A first. After this, we can use the same algorithm recursively for X, Y, Z, U, respectively, and obtain the desired result.

Recall the graduated ruler $\mu_{\ell,j} = (1^{2^j} \cdot 0^{2^j})^{2^{\ell-j-1}}$, ($0 \leq j \leq \ell-1$), which is of length 2^ℓ, begins with 2^j 1's followed by 2^j 0's, and then 1's and 0's alternately. Define

$$\bar{\mu}_{\ell,j} = (0^{2^j}, 1^{2^j})^{2^{\ell-j-1}} = 1^{2^\ell} \oplus \mu_{\ell,j}.$$

(1) First, construct $\mu_{2\ell,\ell-1}$ and $\mu_{2\ell,2\ell-1}$ within $O(\log m) = O(\ell)$ time and $O(m^2) = O(2^{2\ell})$ space.

(2) Suppose that matrix A is initially in V_1. Then $V_1 \wedge \mu_{2\ell,2\ell-1}$ is the

upper half of the matrix while $V_1 \wedge \mu_{2\ell,\ell-1}$ is the left-hand half of the matrix. Therefore

$V_1 \wedge (\mu_{2\ell,\ell-1} \leftrightarrow \mu_{2\ell,2\ell-1})$ is the matrix $\begin{pmatrix} X & 0 \\ 0 & U \end{pmatrix}$

$V_1 \wedge (\bar{\mu}_{2\ell,\ell-1} \wedge \mu_{2\ell,2\ell-1})$ is the matrix $\begin{pmatrix} 0 & Y \\ 0 & 0 \end{pmatrix}$

$V_1 \wedge (\mu_{2\ell,\ell-1} \wedge \bar{\mu}_{2\ell,2\ell-1})$ is the matrix $\begin{pmatrix} 0 & 0 \\ Z & 0 \end{pmatrix}$

Since shifting a distance $2^{2\ell-1}$ to the right moves the upper half to the lower half, while shifting a distance $2^{\ell-1}$ to the left moves the right-hand half to the left-hand half, we see that

$V_1 \wedge (\bar{\mu}_{2\ell,\ell-1} \wedge \mu_{2\ell,2\ell-1}) \downarrow (2^\ell-1)2^{\ell-1}$ is the matrix $\begin{pmatrix} 0 & 0 \\ Y & 0 \end{pmatrix}$

$V_1 \wedge (\mu_{2\ell,\ell-1} \wedge \bar{\mu}_{2\ell,2\ell-1}) \uparrow (2^\ell-1)2^{\ell-1}$ is the matrix $\begin{pmatrix} 0 & Z \\ 0 & 0 \end{pmatrix}$

Thus after executing the following sentence

$V_1 := (V_1 \wedge (\mu_{2\ell,\ell-1} \leftrightarrow \mu_{2\ell,2\ell-1})) \vee$

$(V_1 \wedge (\bar{\mu}_{2\ell,\ell-1} \wedge \mu_{2\ell,2\ell-1}) \downarrow (2^\ell-1)2^{\ell-1}) \vee$

$(V_1 \wedge (\mu_{2\ell,\ell-1} \wedge \bar{\mu}_{2\ell,2\ell-1}) \uparrow (2^\ell-1)2^{\ell-1})$,

the matrix $\begin{pmatrix} X & Z \\ Y & U \end{pmatrix}$ is in V_1.

Now we do the same for X,Y,Z and U. To do so, we need only to modify the shifting distances and the graduated rulers. We therefore obtain the program:

BEGIN
    suppose m = 2^ℓ
    X := $\mu_{2\ell,\ell-1}$
    Y := $\mu_{2\ell,2\ell-1}$
    Z := $+1^\ell 0^{\ell-1}$ $((2^\ell-1)2^{\ell-1})$, shifting distance
    M := $+10^{\ell-1}$ $(2^{\ell-1})$, to modify X

$\quad$ N := +10^{2\ell-1} $\qquad$ ($2^{2\ell-1}$), to modify Y

$\quad$ WHILE M ≠ + DO

$\qquad$ {M := M ↓ 1

$\qquad$ N := N ↓ 1

$\qquad$ V_1 := (V_1 ∧ (X ↔ Y)) ∨ (V_1 ∧ (¬X ∧ Y)) ↓ Z ∨ (V_1 ∧ (X ∧ ¬Y)) ↑ Z

$\qquad$ X := X ⊕ (X ↓ M) $\qquad\qquad$ to modify X

$\qquad$ Y := Y ⊕ (Y ↓ N) $\qquad\qquad$ to modify Y

$\qquad$ Z := Z ↓ 1} $\qquad\qquad\qquad$ to modify Z

$\quad$ END.

$\quad$ To construct the graduated rulers, the program needs time $O(\log m) = O(\ell)$, and space $O(2^{2\ell}) = O(m^2)$.

THEOREM 27.1 There is a vector machine computing a $2^\ell \times 2^\ell$ matrix transpose within time $O(\ell)$ and space $O(2^{2\ell})$.

$\quad$ Notice that this algorithm depends on the fact that the dimension m is a power of 2. See Exercise 27.1.

$\quad$ In the remaining of this section we discuss how to compute the projection of a word by a vector machine.

$\quad$ Suppose that $\{a,b,\phi\}$ is an alphabet and that f is a homomorphism from $\{a,b,\phi\}^*$ to $\{a,b\}^*$ satisfying $f(a) = a$, $f(b) = b$, $f(\phi) = \Lambda$. Then call f a projection from $\{a,b,\phi\}^*$ to $\{a,b\}^*$ (relative to ϕ), call $f(w)$ the projection of word w. For example, if w = aab$\phi\phi$baϕb then $f(w)$ = aabbab.

$\quad$ We use 10 to represent letter a, 11 to represent letter b, and 00 to represent letter ϕ. We call them the codings of a, b, ϕ and denote them by $C(a)$, $C(b)$, $C(\phi)$ respectively. Namely,

$\qquad$ $C(a) = 10$, $C(b) = 11$, $C(\phi) = 00$.

The coding of $w \in \{a,b,\phi\}^*$ is defined by

$\qquad$ $C(\Lambda) = \Lambda$

$\qquad$ $C(wx) = C(w)C(x)$ $\qquad$ $w \in \{a,b,\phi\}^*$, $x \in \{a,b,\phi\}$.

Therefore the coding of $\phi ab \phi a$ is 0010110010. The problem is then to compute

$C(f(w))$ from $C(w)$.

We only consider the case when the length of $C(w)$ is a power of 2, $|C(w)| = 2^m$.

Assume that a register X is divided into 2^d intervals, each of length 2^e, ($e + d = m$). In each interval i, there is a word $w_i \in \{10,11\}^*$. In other words, the contents of interval i is a word from $(00)^* w_i (00)^*$. Can we move all words w_i to the right-hand end of interval i effectively?

Since we need to accomplish the work on all 2^d intervals, we should partition these intervals into two classes. The first class consists of those intervals in which the number of 0's at the right-hand end is $\geq 2^{e-1}$. The second class consists of the others. We move all contents of the intervals of the first class a distance 2^{e-1} to the right, and leave the contents of second class intervals unchanged. Then for each interval the number of 0's at the right-hand end will be $< 2^{e-1}$.

The next step is to partition the intervals into two classes, according to whether the number of 0's at the right-hand end is $\geq 2^{e-2}$, and move all contents of the first class intervals a distance 2^{e-2} to the right. Then for each interval the number of 0's at the right-hand end will be $< 2^{e-2}$, and so on.

Generally, we move the contents of some intervals a distance 2^j to the right so that the number of 0's at the right-hand end is less than 2^j for each interval. After completing this work for $j = e-1, e-2, \ldots, 2, 1$, all words w_i will be moved to the right-hand end of interval i.

Now the problem is how to partition the intervals so that in the first class the number of 0's at the right-hand end is $\geq 2^j$ but in the second class it is $< 2^j$. Consider the following graduated ruler:

$$v_{je}^d = (0^{2^e - 2^j} 1^{2^j})^{2^d} \quad (j < e).$$

An interval belongs to the first class iff $X \wedge v_{je}^d$ are all 0's.

But how can we know this? We can add all bits in this interval of length 2^e to its rightmost bit, by the following program:

```
Y := X ∧ v_je^d
I := -01                (integer -1)
do j times (Z := Y ↑ I; Y := Y v Z; I := I ↑ 1)
```

$$Y := Y \wedge v_{oe}^d.$$

Then $X \wedge v_{je}^d$ are all 0's in an interval, iff Y is 0 at the rightmost bit of the corresponding interval. Now we extend this bit to the whole interval:

   I := +1

   DO e times {Z := Y ↑ I; Y := Y v Z; I := I ↑ 1}.

Therefore if the i-th interval belongs to the first class then the contents of the i-th interval of Y are all 0's. Otherwise the contents of the i-th interval of Y are all 1's. Now we can partition these intervals into two classes and only move all contents of the intervals in the first class:

$$X := (Y \wedge X) \vee (\neg Y \wedge X) \uparrow (-2^j).$$

Notice that the graduated ruler $v_{j-1,e}^d$ can be obtained from v_{je}^d by the following formula:

$$v_{j-1,e}^d = v_{je}^d \wedge v_{je}^d \uparrow (-2^{j-1})$$

while $v_{e-1,e}^d = \bar{\mu}_{m,e-1} = 1^{2^m} \oplus \mu_{m,e-1}$ can be constructed in time $O(m)$, (m = e+d). Therefore the whole program is

   PROCEDURE Moveright(X,m,e)

   BEGIN

     let d=m-e;

     $V := v_{e-1,e}^d$;

     $U := v_{o,e}^d$;

     $J := +1^{e-1}$;

     $K := -010^{e-1}$;

     WHILE   J ≠ + DO

       BEGIN

         Y := X ∧ V;

         I := -01;

```
            DO |J| times (Z := Y ↑ I; Y := Y v Z; I := I ↑ 1);

            Y := Y ∧ U;

            I := +1;

            DO e times (Z := Y ↑ I; Y := Y v Z; I := I ↑ 1);

            X := (Y ∧ X) v (¬Y ∧ X) ↑ K;

            J := J ↓ 1;

            K := K ↓ 1;

            V := V ∧ (V ↑ K)

        END
    END.
```

The WHILE statement will be executed $O(e)$ times, each costs $O(e)$ steps. The total time used is $O(e^2) = O(m^2)$, the space used is $O(2^m)$.

Similarly there is a procedure Moveleft(X,m,e) moving all words to the left-hand end of each interval of length 2^e, as shown by Figure 27.1.

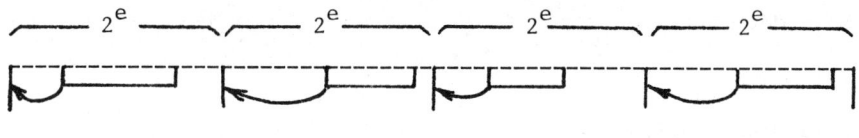

Fig. 27.1

Suppose again that the register X is divided into 2^d intervals, each of length 2^e (e+d=m). At the right-hand end of each interval there is a word w_i. Can we connect the word w_{2i} to the left-hand end of w_{2i-1}? (See Fig. 27.2.)

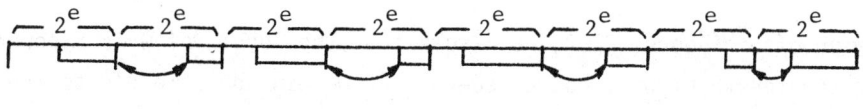

Fig. 27.2

To do so, we need only move w_{2i-1} to the left-hand end of the interval first, and then in all intervals of length 2^{e+1} move all words to the right-hand end. The program is then

```
BEGIN
  Y := Moveleft(X ∧ μ̄_me, m, e)
  X := (X ∧ μ_me) ∨ Y
  X := Moveright(X, m, e+1)
END.
```

Notice that since $\bar{\mu}_{me} = 1^{2^m} \oplus \mu_{me} = (0^{2^e} 1^{2^e})^{2^{d-1}}$, the operation $X \wedge \bar{\mu}_{me}$ is to take out odd intervals of length 2^e. If we execute the above program for $e = 1, 2, \ldots, m-1$, we obtain the projection of X. The whole program is then

```
BEGIN
  FOR e := 1 UNTIL m-1 DO
    BEGIN
      Y := Moveleft(X ∧ μ̄_me, m, e)
      X := (X ∧ μ_me) ∨ Y
      X := Moveright(X, m, e+1)
    END
END.
```

There are altogether $m-1$ iterations. Each iteration costs $O(m^2)$ time. The total time used is $O(m^3)$. The space used is $O(2^m)$. Since the input length $n = |X| = 2^m$, we have

<u>THEOREM 27.2</u> Given the coding $C(w) \in \{00, 01, 11\}^*$ of word $w \in \{a, b, \phi\}^n$, the vector machine can obtain the coding of the projection of w within time $O(\log^3 n)$ and space $O(n)$.

In the proof of Theorem 27.2, we assume that the length of w is a power of 2. If the length of w is not a power of 2, we can add some ϕ's to the left-hand end of w. The time (and space) complexity remains of the same order.

Exercises

27.1 Without assuming that n is a power of 2, prove that the vector machine can compute the transpose of an n × n Boolean matrix, with time $O(\log^2 n)$ and space $O(n^2)$. Is it possible in $O(\log n)$ time and $O(n^2 \log n)$ space?

27.2 Compare Exercise 26.10 and Theorem 27.2. Whey does Exercise 26.10 need only $O(\log^2 n)$ time while Theorem 27.2 needs $O(\log^3 n)$ time?

27.3 Suppose that $w \in \{a,b,\phi\}^*$, $w = w_1 w_2 \ldots w_m$, $|w_i| = \ell$ and $w_i \in \phi^* \{a,b\} \phi^*$ (i.e., there is only one non-ϕ symbol in w_i), $i = 1,2,\ldots,m$. Prove that a vector machine can compute the projection of w in time $O(\log^2(\ell m))$ and space $O(\ell m)$.

27.4 Suppose that $V_1 = P_{11} P_{12} \ldots P_{1n} P_{21} P_{22} \ldots P_{2n} \ldots P_{n1} P_{n2} \ldots P_{nn}$, $V_2 = P_{11} P_{21} \ldots P_{n1} P_{12} P_{22} \ldots P_{n2} \ldots P_{1n} P_{2n} \ldots P_{nn}$, and $|P_{ij}| = r$ (i,j = 1,2,\ldots,n). Design a vector machine transposing V_1 to V_2 in time $O(\log^2(rn))$ and space $O(rn^2)$.

§28 VM SIMULATING LSTM

In order to prove LSTM (phase, width) ≤ *VM (time, space), we prove a theorem announced without proof by Pratt and Stockmeyer (1978): for any deterministic TM T of space complexity S(n), there is a vector machine V simulating T within $O((s(n) + \log n)^3)$ time and $O(n^3 c^{s(n)})$ space, where c is a constant.

In section 19, we introduced the concept of a description word for TM. The description word determines the next movement of the machine and the output symbol. Therefore the action of the TM can be expressed by a directed graph with all possible description words as its nodes. If the machine can reach a description word D_2 from description word D_1 in one step, and outputs symbol a, (possibly Λ), then there is an edge labelled a, from node D_1 to node D_2. This graph has at most $O(n \cdot c^{s(n)})$ nodes. The fan-out of each node is at most 1. (This kind of graph is called a deterministic graph.) The total output of the TM is the assignment of the path of length $t = O(nc^{s(n)})$ starting from the initial description word. Without loss of generality, we may assume that t is a power of 2.

Thus, we consider the following problem.

Suppose that Σ is a finite alphabet, G is a deterministic Σ-assignment directed graph, whose nodes are $d_1, d_2, \ldots, d_t$. Each node has fan-out number 1. Each symbol a in Σ can be represented by a binary string $\bar{a}$ of length e, the coding of a. For simplicity, we may assume $d_1, d_2, \ldots, d_t$ are all binary strings of length ke ($k \le t$). Therefore the determinisitic graph G can be represented by three vectors (the coding of G):

$$d_1 \quad d_2 \quad \ldots \quad d_t$$
$$d_1' \quad d_2' \quad \ldots \quad d_t' \quad\quad (28.1)$$
$$v_1 \quad v_2 \quad \ldots \quad v_t$$

where (d_i, d_i') is the edge starting from node d_i; $v_i = \bar{c}_i \bar{0} \ldots \bar{0}, (\bar{0} = 0^e)$ represents the fact that the assignment on edge (d_i, d_i') is c_i, $|d_i| = |d_i'| = |v_i| = ke$.

There is a unique path $P(d_i)$ of length t starting at node d_i. Our task is to find the end point $\tilde{d}_i$ of path $P(d_i)$ and the assignment of path $P(d_i)$, by a vector machine. For convenience, we shall call a path of length ℓ an ℓ-path.

THEOREM 28.1   There is a vector machine V which

(1) given the coding (28.1) of the graph G and $1^t, 1^k, 1^e$ ($k \le t$), can obtain the t-path table of graph G:

$$d_1 \, 0^{(t-k)e} \quad d_2 \, 0^{(t-k)e} \quad \ldots \quad d_t \, 0^{(t-k)e}$$
$$\tilde{d}_1 \, 0^{(t-k)e} \quad \tilde{d}_2 \, 0^{(t-k)e} \quad \ldots \quad \tilde{d}_t \, 0^{(t-k)e} \quad\quad (28.2a)$$
$$\tilde{v}_1 \quad\quad \tilde{v}_2 \quad\quad \ldots \quad \tilde{v}_t$$

where $\tilde{d}_i$ is the end point of t-path $P(d_i)$, $\tilde{v}_i$ is the assignment of $P(d_i)$ (since $|\tilde{v}_i| = te$, we put some 0's to the right of d_i so that the lengths of d_i and $\tilde{d}_i$ are all te);

(2) uses $O(\log^2(te))$ time and $O(t^3 e)$ space.

Proof: the task can be divided into several parts.

(1) Drill some "holes" into the coding (28.1) as in Example 26.8 to obtain

$$d_1 0^{(t-k)e} \quad d_2 0^{(t-k)e} \quad \ldots \quad d_t 0^{(t-k)e}$$

$$d_1' 0^{(t-k)e} \quad d_2' 0^{(t-k)e} \quad \ldots \quad d_t' 0^{(t-k)e} \qquad (28.2b)$$

$$v_1 0^{(t-k)e} \quad v_2 0^{(t-k)e} \quad \ldots \quad v_t 0^{(t-k)e}$$

This needs $O(\log(tke)) = O(\log(te))$ time and $O(t^2 e)$ space.

The purpose is to leave enough space to store $\tilde{v}_1, \ldots, \tilde{v}_t$. After the work, we have already obtained a 1-path table. In the following we still denote $d_i 0^{(t-k)e}$, $d_i' 0^{(t-k)e}$, $v_i 0^{(t-k)e}$ by d_i, d_i', v_i respectively. But their lengths become te. Set

$$P_i = \begin{pmatrix} d_i \\ d_i' \\ v_i \end{pmatrix}$$

Then we obtain $P_1 P_2 \ldots P_t$.

(2) Obtain the following results on six registers by copying and holing

$$\begin{pmatrix} X_1 \\ X_2 \\ X_3 \end{pmatrix} = \overbrace{P_1 P_1}^{t} \ldots \overbrace{P_1 P_2 P_2}^{t} \ldots P_2 \ldots \overbrace{P_t P_t}^{t} \ldots P_t \qquad (28.3)$$

$$\begin{pmatrix} Y_1 \\ Y_2 \\ Y_3 \end{pmatrix} = \overbrace{P_1 P_2}^{t} \ldots \overbrace{P_t P_1 P_2}^{t} \ldots P_t \ldots \overbrace{P_1 P_2}^{t} \ldots P_t \qquad (28.4)$$

This needs $O(\log(te))$ time and $O(t^3 e)$ space.

(3) Notice that every pair $\begin{pmatrix} P_i \\ P_j \end{pmatrix}$ ($i,j=1,2,\ldots,t$) occurs in the above two rows. If $d_i' = d_j$ in this pair, then (d_i, d_i') and (d_j, d_j') can be connected to form a 2-path. We should produce $p_i' = \begin{pmatrix} d_i \\ d_j' \\ v_i' \end{pmatrix}$ where $v_i' = \bar{c}_i \bar{c}_j \bar{0} \ldots \bar{0}$ representing that there is a 2-path from d_i to d_j' with assignment $c_i c_j$ (i.e., v_i').

How can we tell that $d_i' = d_j$? We know that $d_i' = d_j$ iff $d_i' \leftrightarrow d_j = 1^{te}$. Therefore the machine can execute the following instruction:

$Z := X_2 \leftrightarrow Y_1$; (Z consists of t^2 intervals, each of length $|P_i| = te$. Among every t intervals there is exactly one, whose contents is 1te. This interval is the one we are finding.)

$$Z := \text{Fill}_0(Z, t^2, te) \quad \text{(fill 0's to all other intervals)}$$

$$X_2 := Y_2 \wedge Z;$$

$$X_2 := \text{Fill}(X_2, t, t, te) \text{ (put the end points of 2-paths into } X_2\text{)}$$

$$X_3 := X_3 \wedge Z$$

$$Y_3 := Y_3 \wedge Z$$

$$X_3 := X_3 \vee (Y_3 \downarrow e) \quad \text{(obtain the assignments of 2-paths)}$$

$$X_3 := \text{Fill}(X_3, t, t, te) \text{ (put the assignments of 2-paths into } X_3\text{)}$$

(where Fill_0 and Fill are given by Example 26.11 and Exercise 26.11.)

Then the machine obtains

$$\begin{pmatrix} X_1 \\ X_2 \\ X_3 \end{pmatrix} = \overbrace{P_1'P_1' \ldots P_1'}^{t} \overbrace{P_2'P_2' \ldots P_2'}^{t} \ldots \overbrace{P_t'P_t' \ldots P_t'}^{t} \quad (28.5)$$

where

$$P_i = \begin{pmatrix} d_i \\ d_i'' \\ v_i' \end{pmatrix},$$ representing that the 2-path starting at d_i has end point d_i'' and assignment v_i'.

(4) Transpose (28.5) to obtain

$$\begin{pmatrix} Y_1 \\ Y_2 \\ Y_3 \end{pmatrix} = P_1'P_2' \ldots P_t'P_1'P_2' \ldots P_t' \ldots P_1'P_2' \ldots P_t' \quad (28.6)$$

This needs $O(\log(te))$ time and $O(t^3 e)$ space.

Therefore from (28.3) and (28.4) (which are the copies of 1-path table) to obtain (28.5) and (28.6) (which are the copies of 2-path table), we need

$O(\log(te))$ time and $O(t^3 e)$ space.

In the same way, we can get the copies of 2ℓ-path table from the copies of ℓ-path table within $O(\log(te))$ time and $O(t^3 e)$ space.

Repeat the work for $\ell = 1, 2, 4, \ldots t/2$, we obtain the copies for the t-path table, therefore obtain the t-path table itself. The total time used is $(\log t) \cdot O(\log(te)) = O(\log^2(te))$. The space is obviously $O(t^3 e)$.

In the following we are going to prove Pratt and Stockmeyer's theorem. For simplicity, we can assume that T is a TM, it has a unique accepting state q_1; when entering state q_1, the machine stops dynamically in the meaning that it stays in q_1 for ever and does nothing, therefore the next move function δ is defined everywhere; T has only one work tape which is infinite only to the right; when T starts, the input, output and work tape heads scan the leftmost squares of the corresponding tapes; the output tape head can only move to the right and will never write a space ⌴ on the output tape.

Assume that T has $|Q| = 2^e$ states, $2^e - 1$ work symbols, $|\Sigma| = 2^e - 1$. Therefore any state or symbol can be coded by a binary string of length e. Denote 1^e and 0^e by $\bar{1}$ and $\bar{0}$ respectively. Assume further that the coding of the initial state q_0 is $\bar{q}_0 = \bar{0}$, the coding of the accepting state q_1 is $\bar{q}_1 = \bar{1}$. The coding of a space ⌴ is $\bar{0}$. $\bar{1}$ is not a coding for any symbol in Σ.

Given input $a_1 a_2 \ldots a_n$, the description word of T at any time can be represented by two vectors α and β

$$\alpha = \bar{0}\ \bar{a}_1\ \bar{a}_2 \ldots \bar{a}_i \ldots \bar{a}_n\ \bar{0}\ \bar{q}\ \bar{1}\ \bar{b}_1 \ldots \bar{b}_j \ldots \bar{b}_s \bar{1}$$
$$\beta = \bar{0}\ \bar{0}\ \bar{0} \ldots \bar{1} \ldots \bar{0}\ \bar{0}\ \bar{1}\ \bar{0}\ \bar{0} \ldots \bar{1} \ldots \bar{0}\ \bar{0} \qquad (28.7)$$

where $s = s(n)$. The vector represents the state, the contents on the input tape and the contents on the work tape. The vector β represents the head positions. Thus

$$\alpha \wedge \beta = \bar{0} \ldots \bar{0}\ \bar{a}_i\ \bar{0} \ldots \bar{0}\ \bar{q}\ \bar{0} \ldots \bar{0}\ \bar{b}_j\ \bar{0} \ldots \bar{0}$$

represents exactly the state, the input symbol and work symbol scanned.

The next move function δ can be represented by several vectors. For example, $\delta(q, a, b) = (q', b', c, L, R)$ means, if the state is q, the input symbol scanned is a and the work tape symbol scanned is b, that next time the state should be q', the input head should move one square left, the work tape head should change the symbol b to b' and move one square right, the output head should output a symbol c (and move one square right automatically). We can represent this by four vectors

$$\begin{vmatrix} 0 & a & 0 & q & 0 & b & 0 \\ 0 & a & 0 & q' & 0 & b' & 0 \\ 1 & 0 & 0 & 1 & 0 & 0 & 1 \\ c & 0 & 0 & 0 & 0 & 0 & 0 \end{vmatrix} \quad \begin{array}{l} \text{state and symbols scanned} \\ \text{new state and symbol} \\ \text{the movements of heads} \\ \text{output symbol} \end{array} \quad (28.8)$$

For all possible q,a,b, construct such four vectors. Concatenating them together, obtaining four vectors

$$\begin{aligned} A &= A_1 \, A_2 \, \ldots \, A_m \\ B &= B_1 \, B_2 \, \ldots \, B_m \\ C &= C_1 \, C_2 \, \ldots \, C_m \\ D &= D_1 \, D_2 \, \ldots \, D_m \end{aligned} \quad (28.9)$$

where $m = (2^e)^3$ and

$$\delta_i = \begin{vmatrix} A_i \\ B_i \\ C_i \\ D_i \end{vmatrix} \quad (28.10)$$

is in form (28.8). Notice that sometimes the machine T has no output. In this case we output $\bar{0}$, i.e., a space ⊔. We will remove all $\bar{0}$'s by projection later.

We can design the δ function table carefully so that whenever the work tape head scans a symbol $\bar{1}$ in its description word (α and β) (this means the space is not large enough), the machine stops dynamically.

Since, for any TM, there are only finitely many terms in the δ function table, we can construct the δ function table in $O(1)$ time and $O(1)$ space.

In the following two lemmas, e is a constant, $e = \log |Q|$. The proofs of these lemmas are left as exercises.

<u>LEMMA 28.1</u> There is a vector machine which has the following function:

INPUT $\bar{a}_1\bar{a}_2...\bar{a}_m$, 1^n and 1^s.

OUTPUT All description words with input $a_1a_2...a_n$ and space s (s-description words)

$$D_1D_2...D_t$$

where each D_i consists of two vectors in form (28.7),

$$t = (n+2)2^e(2^e)^s(s+2), \qquad (28.11)$$

and

$$D_1 = \begin{pmatrix} \bar{0} & \bar{a}_1 & \bar{a}_2 & ... & \bar{a}_n & \bar{0} & \bar{q}_0 & \bar{1} & \bar{0} & \bar{0} & ... & \bar{0} & \bar{1} \\ \bar{0} & \bar{1} & \bar{0} & ... & \bar{0} & \bar{0} & \bar{1} & \bar{0} & \bar{1} & \bar{0} & ... & \bar{0} & \bar{0} \end{pmatrix} \qquad (28.12)$$

is the initial description word.

COMPLEXITY: time = $O(s+\log n)$, space = $O((n+s)t)$.

[Hint: by Exercise 26.5, Example 26.7, Example 26.10]

LEMMA 28.2 There is a vector machine which has the following function:

INPUT $\bar{a}_1\bar{a}_2...\bar{a}_n$, 1^n and 1^s.

OUTPUT s-table:

$$\begin{array}{cccc} D_1 & D_2 & ... & D_t \\ D_1' & D_2' & ... & D_t' \\ v_1 & v_2 & ... & v_t \end{array} \qquad (28.13)$$

where $D_1, D_2, ..., D_t$ are all possible s-description words, D_i' is the next s-description word of D_i, v_i is the output symbol when the machine goes from D_i to D_i',

$$t = (n+2)(s+2)(2^e)^{s+1}$$

and D_1 is the initial s-description word.

COMPLEXITY: time = $O((s+\log n)^2)$, space = $O(n^3c^s)$ where c is a constant.
[Hint: Construct the δ-function table first, then use the technique in Theorem 28.1, Lemma 28.1, and Exercise 27.3.]

THEOREM 28.2 (Pratt and Stockmeyer). Suppose that T is a TM of space complexity $s(n)$. Then there is a vector machine V simulating T within time $O((s(n)+\log n)^3)$ and space $O(n^3 c^{s(n)})$. By simulating we mean that the input $a_1 a_2 \ldots a_n$ is a binary word, and if the TM outputs $b_1 b_2 \ldots b_m$ then the vector machine outputs

$$\bar{b}_1 \bar{b}_2 \ldots \bar{b}_m.$$

Proof: the vector machine transforms $a_1 a_2 \ldots a_n$ to $\bar{a}_1 \bar{a}_2 \ldots \bar{a}_n$ first, and then take s to be a natural number.

1. find-s-table: accomplish the work in Lemma 28.2. This needs time $O((s+\log n)^2)$ and space $O(n^3 c^s)$

Notice that the s-table (28.13) is the coding of the graph in Theorem 28.1. This graph has $t = (n+2)(s+2)2^{e(s+1)}$ nodes. e is a constant.

2. find-t-path-table: accomplish the work in Theorem 28.1. This needs time $O((s+\log n)^2)$ and space $O(n^3 c^s)$.

The t-path-table obtained has two possibilities.

(i) halt: the state in $\tilde{D}_1$ (the end point of t-path starting at the initial description word) is the accepting state $q_1 = 1$. Now the path assignment $\tilde{v}_1$ differs from the output of machine T only in that $\tilde{v}_1$ may have more $\bar{0}$'s in it. We should goto 3 to project $\tilde{v}_1$.

(ii) error: in D_1, the work tape head scans a symbol 1. This means the integer s is not large enough. We should enlarge s and try again from the beginning.

3. project-output: remove all $\bar{0}$'s in $\tilde{v}_1$. By Theorem 27.2, the time used is $O(\log^3 t) = O((s(n)+\log n)^3)$, space used is $O(t^3) = O(n^3 c^s)$.

Therefore the whole program of V is then

```
BEGIN
  s := 2
L: find-s-table;
   let t=(n+2)(s+2)2^e(s+1)
   find-t-path-table;
```

IF error THEN (s := 2s; goto L) ELSE project-output

END.

Since s is doubled at most log s(n) times, except the last sentence, the time used is $(\log s(n)) \cdot O((s(n)+\log n)^2)$. The last sentence uses time $O((s(n)+\log n)^3)$. The total time used is $O((s(n)+\log n)^3)$. The space used is $O(t^3) = O(n^3 c^{s(n)})$.

By this theorem we can easily prove the following result;

THEOREM 28.3 Suppose that (f,g) is a nice pair, L is an (f,g)-LSTM. Then there is an (f^*,g^*) vector machine V simulating L.

Proof: since each time the transformation of L is one performed by a TM in space $O(\log g)$, there is a vector machine V simulating L. To simulate each transformation of L, the time used by V is $O((\log g+\log g)^3) = O(\log^3 g)$. The total simulating time is $f \cdot O(\log^3 g) = O(f^*)$. The space used by V is obviously $O(g^3 c^{\log g}) = O(g^*)$.

Exercises

28.1 Prove that any TM of space complexity s(n) can be simulated by a TM with the same space complexity and having only one work tape which is infinite only to the right.

28.2 Prove Lemma 28.1, and Lemma 28.2.

28.3 Prove that in the proof of Theorem 28.2, in order to test whether the case "halt" occurs or the case "error" occurs, the vector machine needs time $O(\log(ns))$.

28.4 Prove that a TM of space complexity $O(\log n)$ can be simulated by a $(\log^3 n, n^*)$ vector machine.

28.5* Prove Theorem 28.1 without assuming that t is a power of 2.

§29 THE SIMILARITY BETWEEN VM, RAM AND TM

In this section it is proved that vector machines can be simultaneously simulated by TM (reversal, space), thereby completing the proof of the similarity of the following models:

TM(reversal, space)

RAM(reversal, space)

VM(time, space).

THEOREM 29.1 Suppose (f,g) is a nice pair, and V is an (f,g)-VM. Then there is an (f^*,g^*) TM T simulating V.

Proof: construct TM T as follows: The work alphabet is $(0,1,+,-,\$,\sqcup)$.

If a vector $+w$ or $-w$ ($w \in \{0,1\}^*$) is in the i-th register, then $+w\$$ or $-w\$$ is in the i-th work tape of T. Notice that the symbol $+$ in the case of a vector machine means a string of 0's infinite to the left, whereas, in the case of our TM, it is just a symbol. The symbol $ represents the right-hand end point of a vector. Usually the tape heads point to the square to the left of the symbol $.

The TM can simulate the following instructions:

  A := vector

  A := B ∧ C

  A := B

and can test predicates $A = +$ or $A \neq +$ in two phases (going from right to left and then coming back to the right).

For instruction A := B ↑ C, if the number(C) > 0 then the TM can transform the number(C), which is in binary form, to unary form within $O(\log g(n))$ = $O(f^*)$ reversal. At the same time it moves the head of tape B a distance of number(C) to the right, and copies the contents of tape B to tape A.

If the number(C) < 0, then the TM generates the unary representation of -number(C) by considering the lower bits first and at the same time moves the head of tape B left. If, when the unary representation of -number(C) has been completely generated, the head of tape B is still in the range of its contents (the head has not scanned a + or a - symbol), it then copies the contents (to the left of the head) of tape B to tape A. If when the head of tape B has passed the whole contents of tape B, the TM is still generating the unary representation of -number(C) then, except the sign + or -, there is nothing left after shifting -number(C) squares to the right. The TM needs only to copy the sign + or - to tape A. The reversal used is

134

$O(\log g(n))$, the space used is $O(g(n))$.

In short, the whole simulating time of V is $O(f \cdot \log g) = O(f^*)$, the space used is $O(g)$.

Thus, in the flat case we have proved the similarity between the models shown in Fig. 29.1

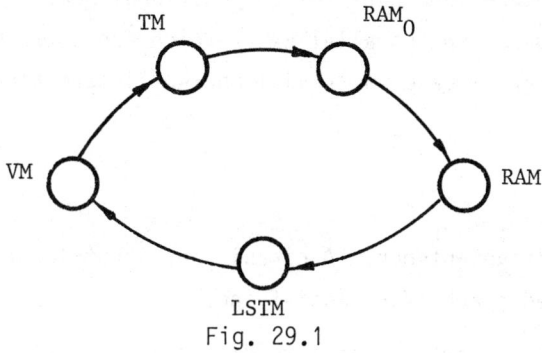

Fig. 29.1

The narrow case will be discussed in Chapter 7, §34, Remark 34.1.

7 Other parallel computational models

In this chapter we introduce some parallel computational models — uniform circuits, uniform aggregates and parallel RAM — which are often used in the literature. Since the narrow case is trivial and not interesting, we only consider the flat case.

§30 UNIFORM CIRCUITS

Suppose that n is a positive integer. A circuit for input length n is a labelled acyclic directed graph (V,E) satisfying:

(1) The vertex set V is partitioned into d+1 levels: level 0, level 1,...,level d. There are exactly n vertices in level 0. They are called the input gates of the circuit. The number d is called the depth. The maximum number of vertices in level 1 through level d is called the width of the circuit.

(2) Each vertex in level 1 through level d is labelled with $\neg$, $\vee$ or $\wedge$ (the Boolean operations). For any vertex v labelled with $\neg$ in level ℓ, $1 \leq \ell \leq d$, there is a unique u in level $\ell-1$ or level 0 such that $(u,v) \in E$ is an edge (the input line). For any vertex v labelled with $\vee$ or $\wedge$ at level ℓ, $1 \leq \ell \leq d$, there are exactly two vertices u_1 and u_2 in level $\ell-1$ or level 0 such that $(u_1,v) \in E$ and $(u_2,v) \in E$ (two input lines).

(3) Suppose that there are $N = |V|$ vertices in V. There is a one to one mapping ϕ from V to the set $\{1,2,...,N\}$ such that if v_1 is in level ℓ_1, v_2 is in level ℓ_2 and $\ell_1 < \ell_2$, then $\phi(v_1) < \phi(v_2)$. The natural number $\phi(v)$ is called the name of the vertex v. The number N is called the size of the circuit.

The input of the circuit is a binary string of length n. Each vertex in V has a value: if the vertex named i is in level 0, then $1 \leq i \leq n$ and its value is the i-th input bit. If the vertex is in level ℓ, $1 \leq \ell \leq d$, then its value is defined by the Boolean operation in the obvious way. The values of some designated vertices in level d, named N-m+1,...,N, are considered to be the results of the computation.

For each non-input gate, we give an information segment

(type, name, input1, input2),

where "name" is the name of the gate, "type" is the Boolean operation of the gate, "input1" and "input2" are the names of the gates the input lines (edges) come from. If the type is $\neg$, then there is only one input. We assume that input1 and input2 are the same. A coding of the circuit is a collection of information segments of all non-input gates, and a natural number m.

<u>DEFINITION 30.1</u> A uniform circuit (UC) is a family of circuits $\{C_n | n=1,2,\ldots\}$ satisfying

(1) the circuit C_n has exactly n input gates and
(2) the coding of C_n is log-space constructible.

If (f,g) is a nice pair and the depth and width of the uniform circuits are not more than f and g respectively, then we call the circuits an (f,g) - UC.

If a circuit has n input gates and m output gates, then it realizes a mapping from n variables to m variables. Or we can say this (n,m) mapping is realized by the circuit.

<u>LEMMA 30.1</u> Suppose that g(n) is log-space constructible. Let p be the following (g(n),1) mapping:

$$p: (a_1, a_2, \ldots, a_{g(n)}) \to \neg (a_1 \vee a_2 \vee \ldots \vee a_{g(n)}).$$

Then p can be realized by a $(1 + \lceil \log g(n) \rceil, g(n))$ uniform circuits.

Proof: the circuits are in fact binary or-trees, with a "$\neg$" gate added to the root. Since g(n) is log-space constructible, so is the family of the codings of the trees. Therefore the circuits are uniform.

<u>LEMMA 30.2</u> Suppose that g(n) is log-space constructible and A,B,C are vectors (refer to Chapter 6) satisfying C = A ↑ B, and that all their lengths |A|, |B|, |C| are bounded by g(n). (Therefore when shifting to the left the shifting distance is bounded by g(n)-|A|.) Then the mapping from the values

of A and B to the values of C can be realized by an $(O(\log g(n)), O(g(n)))$ uniform circuits.

Proof: the circuits are composed of $\ell = \lceil \log g(n) \rceil$ levels, level 0, level 1,...,level $\ell-1$. At level i, the values in vector A are shifted a distance 2^i to the left, to the right, or remain in the same positions, according to the sign and the i-th bit of vector B. The details are left to the reader.

We are going to simulate vector machines by uniform circuits. Suppose that (f,g) is a nice pair of functions and V is an (f,g)-vector machine with k vectors $V_1, V_2, \ldots, V_k$ and q states $Q_1, \ldots, Q_q$. Let $Q_i(t)$ be the proposition that at time t the vector machine V is in state Q_i. Obviously, for each t there is a unique i such that $Q_i(t)$ is true. Let $V_i(j,t)$, ($i = 1,2,\ldots,k$, $j = 1,2,\ldots,g(n)$), be the proposition that the j-th bit of vector V_i at time t is 1. Let $A_i(t)$ be the proposition that at time t the contents of vector V_i is + (i.e., $V_i(1,t), V_i(2,t),\ldots,V_i(g(n),t)$) are all 0's; refer to Chapter 6).

The state of V at time t+1 is completely determined by the state of V at time t and the values $A_1(t), A_2(t), \ldots, A_k(t)$. The mapping

$$\{Q_i(t)|i=1,2,\ldots,q\} \cup \{A_i(t)|i=1,2,\ldots,k\} \to \{Q_i(t+1)|i=1,2,\ldots,q\}$$

can be realized by an $(O(1), O(1))$ circuit, since k and q are all constants. This circuit is a uniform circuit by definition. By Lemma 30.1, the mapping

$$\{Q_i(t)|i=1,2,\ldots,q\} \cup \{V_i(j,t)|j=1,\ldots,g(n), i=1,\ldots,k\} \to \{Q_i(t+1)|i=1,2,\ldots,q\}$$

can be realized by an $(O(\log g(n)), O(g(n)))$ uniform circuit.

If we know the state of V at time t, say it is Q_i, then the mapping ϕ_i is completely determined:

$$\phi_i : \{V_\ell(j,t)|\ell=1,\ldots,k, j=1,\ldots,g(n)\} \to \{V_\ell(j,t+1)|\ell=1,\ldots,k, j=1,\ldots,g(n)\}.$$

It is obtained either by a Boolean operation ($\neg$ or $\wedge$), or by a shifting operation. By Lemma 30.2, the mapping ϕ_i can be realized by an $(O(\log g(n)), O(g(n)))$ uniform circuit. Since the following mapping

$$\phi : \{Q_i|i=1,\ldots,q\} \cup \{V_\ell(j,t)|\ell=1,\ldots,k, j=1,\ldots,g(n)\}$$
$$\to \{V_\ell(j,t+1)|\ell=1,\ldots,k, j=1,\ldots,g(n)\}$$

can be expressed as

$$\phi = Q_1(t)\phi_1 \vee Q_2(t)\phi_2 \vee \ldots \vee Q_q(t)\phi_q,$$

the mapping ϕ can be realized by an $(O(\log q(n)), O(q(n)))$ uniform circuit. Therefore the mapping ψ

$$\psi: \{Q_i(t) | i = 1,\ldots,q\} \cup \{V_i(j,t) | i=1,\ldots,k, j=1,\ldots,g(n)\}$$
$$\rightarrow \{Q_i(t+1) | i=1,\ldots,q\} \cup \{V_i(j,t+1) | i=1,\ldots,k, j=1,\ldots,g(n)\}$$

can be realized by an $(O(\log g(n)), O(g(n)))$ uniform circuit.

That is to say, one step of the vector machine can be simulated by an $(O(\log g(n)), O(g(n)))$ uniform circuit. By Theorem 19.3, the vector machine V can be simulated by an $(O(f \cdot \log g(n)), O(g(n)))$ uniform circuit. Thus we have the following theorem.

THEOREM 30.1 Suppose (f,g) is a nice pair. Then an (f,g) vector machine can be simulated by an $(O(f \cdot \log g(n)), O(g(n)))$ uniform circuit.

The following theorem is a very important and interesting result.

THEOREM 30.2 (Borodin) An (f,g) uniform circuit can be simulated by a TM of space complexity $O(f \log(fg))$.

Proof: since the coding of the circuit C is log-space constructible, the TM can find the information segment (type, name, input1, input2) for any given gate "name", in $O(\log(fg))$ space. Thus we can compute the value of "name" by the following procedure:

    PROCEDURE VALUE (name)
    IF name $\leq$ n THEN return the name-th bit of the input ELSE
    {(type, name, input1, input2) := information segment of name;
    if type = $\neg$ then return ($\neg$ Value(input1));
    if type = V then return (Value(input1)$\vee$Value(input2));
    if type = $\wedge$ then return (Value(input1)$\wedge$Value(input2));}

The recursive depth is $O(f)$. At each level it needs $O(\log(fg))$ space.

In $O(\log(fg))$ space, the TM can compute the natural number m (which is the total number of output gates) and the largest name N in the circuit. Therefore the following program can obtain the output of the circuit in $O(f \cdot \log(fg))$ space. The program is:

   FOR i = N-m+1 upto N DO output Value(i);

Since an (f,g) VM can be simulated by an (f^*,g^*) UC, the vector machine can be simulated by a $(2^{f^*},f^*)$ TM according to the above theorem. This is shown in Fig. 30.1. This result is somehow dual to Theorem 28.2, which is shown in Fig. 30.2. The flat vector machine can be transformed into a narrow TM, and the narrow TM can be transformed into a flat vector machine.

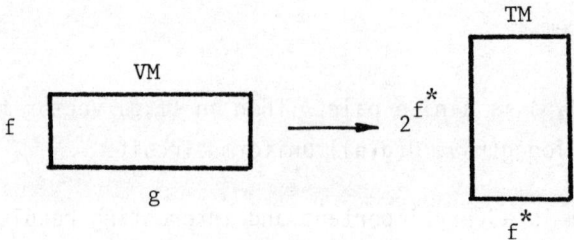

Fig. 30.1

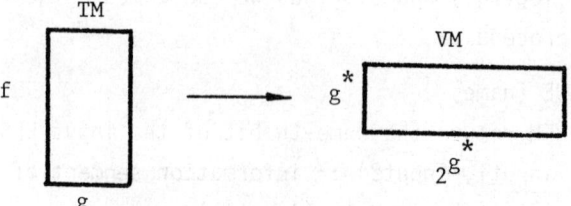

Fig. 30.2

Exercises

30.1 Prove that if T is a TM of space complexity $O(\log n)$, then there is a $(\log^* n, n^*)$ TM simulating T.

30.2 Prove that a transformation class belongs to NC iff it can be realized by a $(\log^* n, n^*)$ TM.

30.3 Prove that in Lemma 30.1, the coding of the trees is log-space constructible.

30.4 Suppose that (f,g) is a nice pair, h is log-space constructible and we have $f^* g^* \geq h$. Then the unary representation of $h(n)$ can be computed within f^* reversal and g^* space by a TM.

30.5 Suppose that (f,g) is a nice pair, $\{C_n\}$ is a log-space constructible coding, and we have $|C_n| \leq f^*(n) g^*(n)$. Then C_n can be obtained within f^* reversal and g^* space, given 1^n as input.

30.6 (The prefix algorithm) Suppose that there is an associative operation. We want to compute the product of all prefixes of the sequence $x_1, x_2, x_3, \ldots, x_n$ (i.e., output $x_1, x_1 x_2, x_1 x_2 x_3, \ldots, x_1 x_2 \ldots x_n$). Prove that these n products can be computed by a circuit of depth $\log n$ and size $O(n)$. Each node in the circuit is able to compute the product of two elements. [Hint: consider when n = 8 the network shown in Fig. 30.3.]

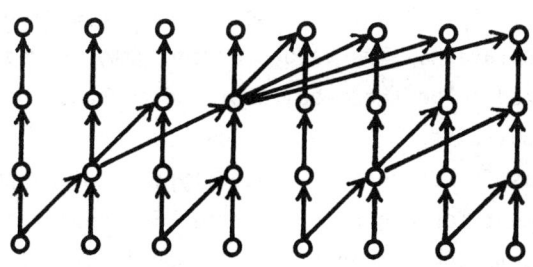

Fig. 30.3

Prove that the coding of the graph is log-space constructible.

30.7* (Pippenger) Prove that an oblivious one-directional (f,g) TM can be simulated by an (f*,g*) uniform circuit.

[Hint: for oblivious TM, we can view each step as a transformation from states to states: $Q \to Q$. This transformation is associative. Therefore we can use the prefix algorithm to simulate it.]

30.8 Prove that an (f,g)-UC can be simulated by a (3f,g)-UC with no V-gates.

§31 UNIFORM AGGREGATES

An aggregate for input length n is a labelled directed graph (V,E) of fan-in 2 satisfying

(1) There are n vertices whose fan-in are 0 (input gates).

(2) Any other vertex (non-input gate) is labelled with $\neg$, $\wedge$ or $\vee$ (the Boolean operations). If it is labelled with $\neg$, then its fan-in number is 1. If it is labelled with $\wedge$ or $\vee$, then its fan-in number is 2.

(3) Suppose that there are $N = |V|$ vertices in the graph. There is a one to one mapping ϕ from V to the set $\{1,2,...,N\}$ such that the input gates are mapped onto $\{1,2,...,n\}$. We call $\phi(v)$ the name of v. The number N is called the space of the aggregate.

At any time $t = 0,1,2,...$, each gate has a value. At time $t = 0$, the value of the input gate named i is the i-th input bit (0 or 1). The value of the non-input gate at time $t = 0$ is 0. The values of all gates at time t is determined by the values of all gates at time t-1 and the Boolean operations in the obvious way. There is a distinguished gate, whose value is initially 0. The time t_0, when the value of this gate becomes 1, is called the time of the computation. The values of the gates named N-m+1,...,N at time t_0 are considered to be the results of the computation, where m is the total number of output bits.

In the same way as for circuits, we use (type, name, input1, input2) as the information segment for a non-input gate. The coding of an aggregate consists of the information segments for all non-input gates and a natural number m.

<u>DEFINITION 31.1</u> A uniform aggregate (UA) is a family of aggregates

$$\{A_n | n = 1,2,3,...\}$$

satisfying

(1) The aggregate A_n has n input gates.
(2) The coding of A_n is log-space constructible.

If (f,g) is a nice pair and the time and space of the uniform aggregates are bounded by f and g respectively, then we call A_n an (f,g)-uniform aggregate.

THEOREM 31.1 In the flat case the uniform circuits can be simulated by the uniform aggregates.

Proof: in fact an (f,g) uniform circuit is an (f,f·(g+n)) uniform aggregate by definition. Since $f \leq g^*$, we have $f \cdot (g+n) \leq f \cdot (g+f^*g^*) \leq g^*$. Therefore the UC can be simulated by UA.

Exercises

31.1 Design the shifting circuits in Lemma 30.2. Prove it is log-space constructible.

31.2 Look at the circuit in Fig. 31.1. In what cases can it "remember" some information?

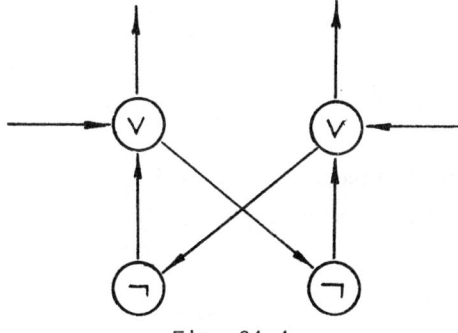

Fig. 31.1

31.3* Can you design an aggregate with two inputs A, B and one output C, such that, whenever the value of A is 1, the aggregate will "remember" and output (after a certain delay) the value of B at this moment, until the next time the value of A becomes 1 again?

31.4 Can you design an FA to accept $L = \{w \in \{0,1\}^* | w$ has even 0's and even 1's$\}$ by means of some $\neg$, $\vee$ and $\wedge$ gates? (Some delay on the input is allowed: i.e., you can input $\phi(w)$ instead of w, where

$$\phi(0) = 0^k, \quad \phi(1) = 10^{k-1},$$

$$\phi(wa) = \phi(w)\phi(a), \quad w \in \{0,1\}^*, a \in \{0,1\}.)$$

§32 PRAM

The parallel random access machines (PRAM) are widely used as a parallel computational model. The main resources considered in most literature are the parallel time and the number of processors used. Do these resources correspond to the reversal and the space of other models? It turns out that the parallel time and space of a PRAM are still the main resources that correspond to the reversal and the space of other models, and the number of processors is not very important as far as polynomial simulation is concerned.

Now we give a definition of the PRAM. Roughly, a PRAM consists of many processors and a common random access memory. Each processor has its own program and works on the common memory in the same way as if it were an RAM. The processors exchange information through the common memory.

More precisely, a PRAM consists of the following two parts:

(1) A set of registers

$$R_0, R_1, R_2, \ldots$$

Each register R_i can hold an integer, content(R_i).
The total number r of used registers may depend on the input length.

(2) A set of processors

$$P_1, P_2, \ldots$$

The total number p of processors may depend on the input length.
Each processor P_j has a name j and a program $PROG_j$.
 A program is described as follows:
 (a) Address:
 direct address R_i
 indirect address $\#R_i$, indicating R_m, where m = content(R_i)

(b) Instruction: Let A, B, C be some addresses and let a be a natural number. Then we have the following three kinds of instructions:

  A := a, put the natural number a into A;

  A := B, put the number content(B) into A;

  A := B*C, put the number content(B)*content(C) into A,

where * is a mapping in NC (admissible instructions, see §21).

Sometimes we restrain the model to have only one kind of instruction A := a, a ∈ {0,1}, and to have only direct addresses. We call this $PRAM_0$.

(c) The program is a directed graph. The vertices in this graph are called the states. The fan-out of every vertex in this graph is less than or equal to 2. If the fan-out of a vertex is 1, then the unique edge out from it is labelled with an admissible instruction; if the fan-out of a vertex is 2, then these two edges are labelled with two opposite predicates A = 0 or A ≠ 0, where A is an address; if the fan-out is 0, then the vertex is a terminating state. A specified state q_0 is called the initial state of the program. All processors may have the same program, but start from different initial states. In any case, the total number of states in the program(s) is $O(p)$.

The input form is the same as for RAM. When comparing with other models, we mainly use the first input form, i.e. the input is a binary string of length n, stored in $R_1, R_2, \ldots, R_n$. In R_0 and R_{n+1}, the integer n is stored.

All the processors start from their own initial states of their program(s). Whenever a processor is in a state of fan-out 1, it performs the corresponding admissible instruction and enters its next state. Whenever a processor is in a state of fan-out 2, it takes the branch on which the corresponding predicate is satisfied. If a processor enters a state of fan-out 0, the processor halts. When all processors halt, the contents of some designated registers are considered to be the output.

We assume that the execution of each instruction or the testing of each predicate needs a unit time and that all the processors work synchronously.

In any case, we shall not allow both reading and writing in any register at the same time by different processors. But sometimes we allow concurrent reading (CR), i.e., more than one processor can read the same register at the same time. Sometimes we allow concurrent writing (CW), i.e., when there are more than one processors writing into one register, the one with the smallest name is the only successful processor. Therefore we have altogether four cases: CRCW, CREW, ERCW, EREW. Here the letter E stands for "exclusive", C stands for "concurrent".

The number $s = r+p$ (the number of registers + the number or processors) is the space. The total number of steps from the beginning to the time when every processor halts is the parallel time t.

Suppose that (f,g) is a nice pair of functions of n. If the following restrictions hold, then we call the machine an (f,g) PRAM (with $\leq$ h processors):

(1) t and s are bounded by f and g respectively.

(2) The bounding function h(n) of the number of the processors is a nice function (ref. Ex. 19.5) and is $\leq g(n)$.

(3) The coding of the program(s) of all processors P_j (j = 1,2,...,p) (including the initial states for all P_j (j = 1,2,...,p)), is log-space constructible, given 1^n as input.

(4) Each register can hold an input of length not more than $O(g^*)$. This restriction is essential, because if the register can hold an arbitrarily long integer, then we cannot compare the PRAM model with others.

(5) $(hf)^* \geq g$. Since, within time t, all p processors can address not more than pt registers, we have $(pt)^* \geq s$ and this restriction is very natural (see Exercise 32.3).

EXAMPLE 32.1 (Merging) There are two sorted sequences of numbers:

$$x_1 \leq x_2 \leq x_3 \leq x_4 \leq \ldots \leq x_n \text{ and } y_1 \leq y_2 \leq y_3 \leq y_4 \leq \ldots \leq y_m, m \leq n.$$

We have p processors. How can we merge them quickly? Using one processor, we can find the place in the sequence of x's where a number, say y_1, should be inserted, within O(log n) steps. Therefore using p processors we can find the corresponding places in the sequence of x's, where the numbers $y_{[m/p]}, y_{[2m/p]}, \ldots, y_m$ should be inserted, within O(log n) steps. For the same reason, we can find the corresponding places in the sequence of y's where the number $x_{[n/p]}, x_{[2n/p]}, \ldots, x_n$ should be inserted. Now these two sequences are divided into not more than 2p segments, each of which is of length $\leq \lceil n/p \rceil$. We can merge the corresponding pairs in parallel by the p processors within time $(O(\lceil n/p \rceil))$. The total time used is $O(\lceil n/p \rceil + \log n)$.

Notice that the lengths of numbers $x_1, x_2, \ldots, y_1, y_2, \ldots$ are all bounded by n^*, and n is the problem size. See Remark 34.5.

Exercises

32.1 Show that sorting n numbers by p processors with $p \leq n$ can be done in $O(n \cdot \log n)/p + \log^2 n)$ time by CRCW-PRAM.

32.2* (Max) Show that there is a CRCW-PRAM to find the maximum element among n elements in time $O(n/p + \log\log p)$.

32.3* Suppose that (f,g) is a nice pair and M is an (f,g) PRAM with $\leq h$ processors, and the maximum length of the integers stored in the registers is bounded by $(fh)^*$. Show that (f,fh) is a nice pair and M can be simulated by an $(f^*, (fh)^*)$ PRAM with $\leq h$ processors.

32.4 Sometimes people define the PRAM in the following way. Each processor P_j has a common program and a private register where the integer j is stored. Prove that this is a special case of our definition.

§33 THE SIMILARITY BETWEEN PRAM, UA, UC AND OTHER MODELS

We have already seen that UA can simulate UC and UC can simulate VM. In this section we shall prove that UA can be simulated by PRAM, which in turn can be simulated by RAM. Thus all these models are similar to multitape Turing machines.

Suppose that there is a nice pair (f,g), and an (f,g) PRAM M. To simulate M by an RAM M', we use registers $R'_0, R'_1, \ldots, R'_{n+1}$ (of M') to store the input; use $R'_{n+5}, \ldots, R'_{n+k}$ to store the program(s) of the processors of M, where k = 5q + 4 and q is the total number of states in all program(s), $q = O(p) = O(s)$; use R'_{n+k+3i} to store the state of the i-th processor of M; use $R'_{n+k+3i+1}$ and $R'_{n+k+3i+2}$ to simulate the register R_{n+i} of M.

Since the coding of the program(s) of the processors of M is log-space constructible, it can be generated by a TM within f^* reversal and g^* space (ref. Exercises 30.4, 30.5). Because TM is similar to RAM, we can first generate the following information (the program(s) and initial states of all processors of M) by the RAM within f^* reversal and g^* space:

(1) If the i-th state is of fan-out 1, then there is an instruction

labelled on the edge out from this state, and the corresponding operation, the addresses of the operation, and the next state are stored in $R'_{n+5i},\ldots,R'_{n+5i+4}$.

(2) If the i-th state is of fan-out 2, then there are two opposite predicates labelled on the two edges out from this state, and a special symbol indicating that this is a predicate, the address of the corresponding predicate and the two next states are stored in $R'_{n+5i},\ldots,R'_{n+5i+4}$.

(3) If the i-th state is of fan-out 0, then a terminating symbol is stored in R'_{n+5i}.

(4) For each processor P_i, $1 \leq i \leq p$, the initial state is stored in R'_{n+k+3i}.

Then, M' simulates M according to the following algorithm:

FOR all j = 1 up to f(n) DO

{FOR all i=h(n) down to 1 DO

{ find the current state q of P_i from R'_{n+k+3i};

find the instruction or predicate of state q from $R'_{n+5q},\ldots,R_{n+5q+4}$;

execute this instruction or test this predicate;

(when we need to read information from R_{n+i} of M, we read the contents in $R'_{n+k+3i+1}$, when we need to write information into R_{n+i} we send the information to $R'_{n+k+3i+2}$, temporarily)

send the next state to R'_{n+k+3i};}

(if more than one processor writes into the same register, then the one with smallest name will succeed)

FOR all i = 1 upto g(n) DO

$R'_{n+k+3i+1} := R'_{n+k+3i+2}$}.

In order to perform the above algorithm, we need several index registers. These do not require extra space, nor does their reuse increase the reversal.

Thus, to simulate one step of the PRAM, the RAM needs only two phases.

THEOREM 33.1 Suppose that (f,g) is a nice pair of functions. Then an (f,g)

(CRCW) PRAM can be simulated by an RAM within f^* reversal and g^* space (both uniform cost and logarithmic cost).

To simulate a UA by a $PRAM_0$ is quite straightforward. If (f,g) is a nice pair of functions, then each gate G_i, $i = 1,\ldots,g(n)$, can be simulated by a processor P_i of the PRAM. We use register R_i to store the value of gate G_i. Without loss of generality, we can assume that the name of the distinguished gate is G_{n+1} and initially the value in R_{n+1} is 0. The program of P_i is very simple:

(1) If the type of G_i is "not" and the input of G_i is G_j, then the corresponding program of P_i is as shown in Fig. 33.1.

WHILE $R_{n+1} = 0$ DO $R_i := \neg R_j$

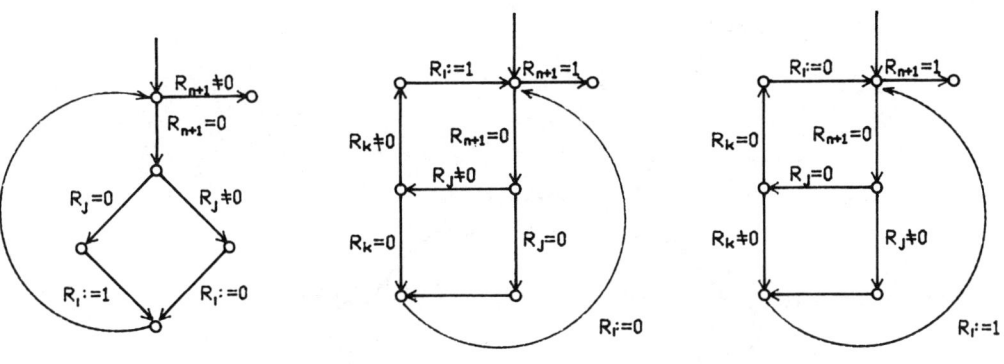

Fig. 33.1   Fig. 33.2

(2) If the type of G_i is "∧" or "∨", and the two input gates are G_j and G_k, then the corresponding program of P_i is

WHILE $R_{n+1} = 0$ DO $G_i := G_j \wedge G_k$

or

WHILE $G_{n+1} = 0$ DO $G_i := G_j \vee G_k$

respectively (Fig. 33.2).

Notice that we have to make the simulation synchronous, thus each step of the UA is simulated by the PRAM with exact four steps. Since the uniform aggregates has unbounded fan-out, the simulating PRAM is of type CREW and is in fact a $PRAM_o$.

Because the coding of the UA is log-space constructible, so is the coding of the simulating program(s) of the PRAM.

THEOREM 33.2 Suppose that (f,g) is a nice pair of functions. Then an (f,g) uniform aggregates can be simulated by an $(O(f),O(g))$ CREW $PRAM_o$.

Combining Theorems 30.1, 31.1, 33.1 and 33.2, we have the following theorem, which is the main result of this chapter:

THEOREM 33.3 The following computational models are similar (see Fig. 33.3):

(1) Multitape TM (reversal space);
(2) PRAM (time, space);
(3) UA (time, space);
(4) UC (depth, width).

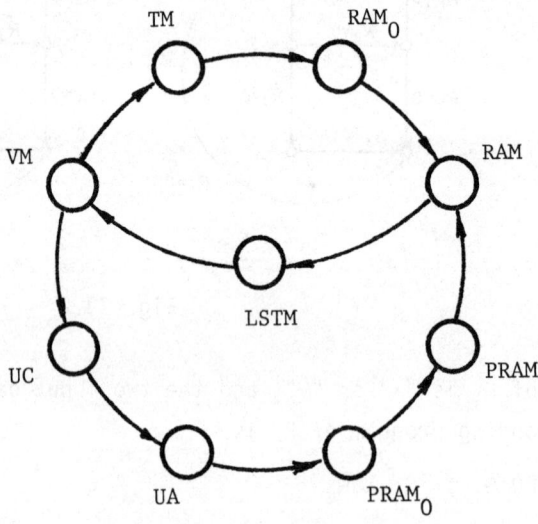

Fig. 33.3

Exercise

33.1 Prove that an (f,g) CREW PRAM can be simulated by an $(O(f \cdot \log g), O(g^2))$ EREW PRAM.

§34 SOME REMARKS ON SIMILARITY

REMARK 34.1 The narrow case. We have discussed the flat case. For narrow case, i.e. when $f^* \geq g$, we can use the traditional method to prove that any two computational models can simulate each other such that the sequential time is polynomially related. The space is also polynomially related. In Chapter 1 of their famous book "The Design and Analysis of Computer Algorithms", Aho, Hopcroft and Ullman proved that the RAM and RASP (random access stored program machine) can be simulated by TM within polynomially related time and, in fact, polynomially related space, provided we use the logarithmic cost criterion.

Since, in the narrow case, the reversal complexity is polynomially related to the sequential time, we conclude that for the narrow case all the models in this book are similar.

What should be noticed is that in the narrow case it may happen that the space bounding function g(n) is much smaller than the input length n. We do not count the input length and output length as work space for TM's. Therefore for other models, we have to use indirect read and write devices.

For aggregates, there is an indirect read device (which consists of many gates) and an input value gate. When the contents of the indirect read device at time t represent a natural number i, the i-th bit of the input word will appear at the input value gate at time t+1. There are two output gates u and v. Whenever the value of u becomes 1, the value of v is considered to be the next output bit. The vector machines can be treated similarly.

For RAM and PRAM, we should assume some input registers and output registers that are not included in the work space.

REMARK 34.2 Sequential time. We have mainly considered parallel time and space. Why not consider sequential time as a main resource? The reason is simple: whenever the parallel time and space complexities of two machines are polynomially related, so is the sequential time.

In fact, for all computational models in this book we always have $t \leq r^* s^*$,

$r \leq t^*$ and $s \leq t^*$. Therefore it is natural to give the following definition:

DEFINITION 34.1 If

(1) (f,g) is a nice pair,
(2) h is a non-decreasing and log-space constructible function,
(3) $h \leq f^*g^*$, and
(4) $f \leq h^*$, $g \leq h^*$,

then we call (f,g,h) a nice triple of functions. If the reversal, space and sequential time complexities of a machine are bounded by f,g,h respectively, where (f,g,h) is a nice triple, then we say the machine is an (f,g,h) machine.

THEOREM 34.1 Suppose that (f,g,h) is a nice triple and M is an (f,g,h) machine in one of the six models discussed in this book. Then in any other model discussed in this book, there is an (f^*,g^*,h^*) machine M_1 simulating it.

Proof: by our similarity theorem there is an (f^*, g^*) machine M_1 simulating M. The sequential time of M_1 is not more than

$$(f^*)^* \cdot (g^*)^* \leq f^*g^* \leq h^*h^* \leq h^*.$$

REMARK 34.3 The number of processors. We have argued that parallel time (reversal) and space are two basic resources for any computational model. Sequential time is polynomially related to the product of reversal and space. Now the reader may ask what corresponds to the number of processors. In the PRAM model, it is a part of space. The total space can be divided into two parts, one is "active" (the processors), the other is "passive" (the registers). Should we define the corresponding resources — the active space — in other models? It turns out that this is not necessary, as far as polynomial simulation is concerned.

Let us first give some definition about bounding functions.

DEFINITION 34.2 Two functions $f(n)$ and $g(n)$ are polynomially comparable if, for any positive integers m_1 and m_2, either

$$f(n)^{m_1} \geq g(n)^{m_2}$$ when n is sufficiently large, or

$$f(n)^{m_1} \leq g(n)^{m_2}$$ when n is sufficiently large.

In fact, almost all pairs from the family of bounding functions used in practice or literature, are polynomially comparable.

LEMMA 34.1 If (f,g) is a nice pair, (f,h), (g,h) are both polynomially comparable pairs, and f,g,h are the bounding functions for parallel time, space and processors of a PRAM, then either

(1) $h \geq^* g$ or

(2) $f \geq^* h$.

Proof: if there is an integer m such that $h^m(n) \geq g(n)$ holds for large enough n, then (1) holds. Otherwise, for any integer m, $h^m(n) \leq g(n)$ holds for large enough n, since h and g are polynomially comparable.

Suppose that condition (2) is not true. Then for any integer m, $f^m(n) \leq h(n)$ holds for large enough n. Thus, for any integer m, $(f(n)h(n))^m \leq h(n)g(n) \leq g(n)^2$ for large enough n. This contradicts the condition that $(fh)^* \geq g$.

In case (1), the total number of active space is polynomially related to the whole space. This is just the case for all parallel models we have mentioned, and just the case for all sequential models we have discussed, because in each reversal the active part may equal the whole space. Thus, in this case, we can assume $h(n) = g(n)$.

In case (2), the PRAM M can be simulated by an RAM within $(fh)^* \leq f*$ sequential time and g^* space. Therefore there is a PRAM M' to simulate M. The machine M' is of f^* time, g^* space and has only one processor, which simulates the processors of M sequentially, one after another. Thus, in case (2), we can assume $h(n) = 1$ as far as polynomial simulation is concerned.

REMARK 34.4 Real computational models. In many applications, we deal with real numbers. We assume that the exact value of the arithmetic operation of two real numbers can be obtained in one step. This results in the following real computational models.

A real multi-index RAM differs form the model RAM_0 only in the following points:

1. An index register can store an integer, but an ordinary register can store an arbitrary real number. If a result is sent to an index register, only the integer part is kept.

2. The admissible instructions are $+$, $\dot{-}$, $*$, $/$ of real numbers and taking the integer part.

3. The space complexity is the total number of ordinary registers used during the computation. If its bounding function is $g(n)$, then the total length of index registers is $O(\log g(n))$.

4. The inputs are n real numbers, $a_1,\ldots,a_n$. Initially, $n,a_1,\ldots,a_n$ are stored in $R_0,R_1,\ldots,R_n$, respectively. The sequential time complexity is the total number of instructions executed and predicates tested.

The real PRAM model differs from PRAM only in that each register can store a real number, the admissible instructions are $+$, $\dot{-}$, $*$, $/$ of real numbers and taking the integer part, and only natural numbers can be used in indirect addressing.

A real PRAM can be simultaneously simulated by a real RAM. The proof is exactly the same as in Theorem 33.1. Therefore we have

THEOREM 34.2 Suppose that (f,g) is a nice pair. Then an (f,g) real PRAM can be simulated by an (f^*,g^*) real RAM.

Conversely, a real RAM can be simulated by a real PRAM. The proof is just like the proofs of Theorem 24.1 and Theorem 28.2. The difference is that when we simulate an RAM by an LSTM, the information is written on the tapes. But when we simulate a real RAM by a real PRAM, the information is stored in registers. The details are left to the reader.

THEOREM 34.3 Suppose that (f,g) is a nice pair. Then an (f,g) real RAM can be simulated by an (f^*,g^*) real PRAM.

Thus these two real computational models are similar. In fact, for every computational model discussed in this book, we can define a corresponding real model. All of them are similar.

Furthermore, we need not restrict ourselves to the real field, we can discuss a general algebraic system, and thereby discuss the similarity of algebraic computational models.

REMARK 34.5 The problem size. Till now we have taken the binary input length of the problem as the problem size. In some cases, for example RAM and PRAM, the input length n is given in some register, but itself is not counted in the input length. Therefore for essentially the same input information we have the same input length in various models.

Generally we can take some other quantity as the problem size, as long as it is polynomially related to the input length and all the models use the same criterion, i.e., in all models we have the same problem size for the same problem. For example, if we want to sort n numbers $x_1, x_2, \ldots, x_n$, each being of length $O(n^*)$, and we use n as the problem size, then we have to use n as the problem size for TM, encoding $x_1, x_2, \ldots, x_n$ as the input by a binary string which is much longer than n. The simultaneous polynomial simulation still holds.

REMARK 34.6 The Multivariable case. We have discussed the case where there is only one input. If there are many inputs, for example two input words x and y with lengths n and m respectively, there is no essential difficulty in getting a similar result. We should define a nice pair of functions $f(n,m)$ and $g(n,m)$ such that

(1) f and g are non-decreasing and log-space constructible, ($f(n,m)$ is log-space constructible if there is an LSTM of phase 1 transforming $1^n 0 1^m$ to $1^{f(n,m)}$);

(2) either $f^* \geq g$ or $g^* \geq f$;

(3) $f^* g^* \geq n+m$;

(4) $f \leq 2^{g^*}$, $g \leq 2^{f^*}$

Then all the proofs are correct.

REMARK 34.7 The constructible character. In our theorems, we always say "for any machine M in Model there is a machine M_1 in $Model_1$ to simulate M such that ...". In fact we can construct such a machine M_1 when machine M is given.

We give each machine in Model a coding, and each maching in $Model_1$ a coding. In fact our theorem is: "There is a TM T, which can transfer the coding of any machine M in Model to the coding of a machine M_1 in $Model_1$ which can simulate M such that ..."

The TM T halts for every input. In fact the computational complexity of T is not high. The exact parallel time, space and sequential time complexities for T present another interesting problem.

Exercises

34.1 We have defined the uniform circuit model. Similarly we can define uniform arithmetic circuits. To do so, we need only make the following changes: the value of each gate (including the input gate) is a natural number, the type of the gates are +, -, *, /, where / is the integer division.
The depth is defined as before; the width of each level is defined to be the sum of the length of the values (integers) of all gates in the level; the width of the circuit is the maximum width among level 1 through level d.
Prove that the uniform arithmetic circuit model is similar to other models.

34.2 The VLSI model is defined by Brent and Kung (1980) and Thompson (1979). Roughly, a VLSI is a layout of an aggregate on a plane grid such that no two gates are laid on the same grid point and no two input lines are laid on the same grid edge (input lines can be laid on paths consisting of many grid edges, and can cross each other). A uniform VLSI is a family of VLSI $\{V_n | n = 1,2,...\}$, whose coding is log-space constructible. The "area" of the VLSI is the area of the rectangle which contains the layout.
Prove that the model Uniform VLSI (time, area) is similar to other models.

34.3 A hardware modification machine is a cell connected to an input tree. The cell consists of a finite control and k input lines. This cell can split and develop into many identical cells connected to each other by their input lines. At any time a cell is in a state $q \in Q$, where Q is a finite state set. At any synchronous step, every cell determines the following actions according to its current state and

the current states of its k neighbours:
(1) it changes its state into another state;
(2) it reconnects its i-th input line from its current i-th neighbour to some neighbour of the i-th neighbour;
(3) if its i-th input line is connected to itself, it can then produce a son, connect its i-th input line to the son, put the son in some state and connect all the input lines of the son to the father.

Obviously, for the same i, the cell can do action (2) or (3), but not both.

We use Q to represent the finite state set. Then the state of the cell and the states of its k neighbours can be represented by an element in Q^{k+1}. Let $K = \{1,2,\ldots,k\}$ be a set. Then the first action can be represented by a mapping $Q^{k+1} \to Q$. The second and third action for the i-th input line can be represented by a mapping $Q^{k+1} \to K \cup Q$. Therefore a HMM can be represented by a mapping $Q^{k+1} \to Q \times (K \cup Q)^k$. In Q, there are a starting state q_0, an accepting state q_1 and a rejecting state q_2. When it begins, the machine has only one cell which is in state q_0. When this cell enters q_1 or q_2, the whole machine stops, accepting or rejecting the input.

The input bits are placed on the leaves of a tree (we use the state of the leaf to represent 0 or 1). Every node of the tree is an identical cell of the HMM. But they are stationary, i.e., their connection cannot be changed, as shown by Fig. 34.1. Only the cell connected to the root can be developed into a group of cells. The hardware is the total number of developed cells excluding the cells on the input tree. The time is the total number of steps.

Prove that the HMM (time, hardware) model is similar to other models.

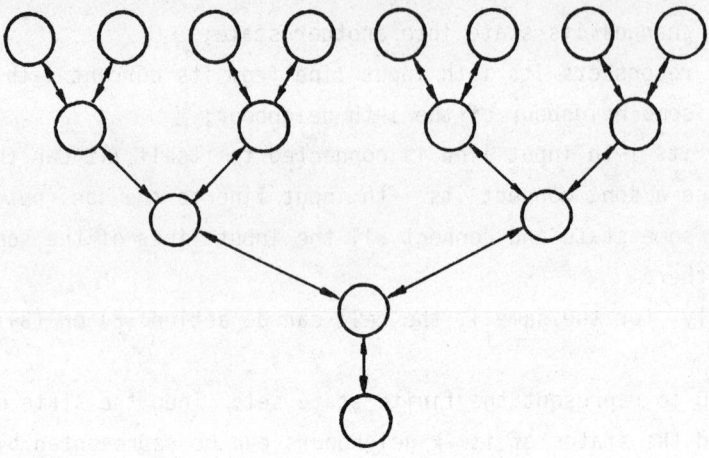

Fig. 34.1

34.4 The storage modification machine (SMM) has a finite control, a read-only tape (tape 0) and k work tapes (tape 1,2,...,k), just like a multitape TM. Each tape cell has two heads (L and R) connecting to its left and/or right-hand neighbours. According to its state and the symbols of the cells that its heads are pointing to, the machine can (1) change to a new state, (2) write down a symbol on the cell that its i-th work tape head is pointing to, (3) move its (input and work) tape heads from cells to their neighbours (we use L, R, S to denote the directions of the movements), and (4) move the L, R heads of the cell that the i-th (i = 1,2,...,k) head of the finite control is pointing to, to the cell that the j-th (j = 1,2,...,k) head of the finite control is pointing to.

Let $K = \{1,2,...,k\}$ be the set of work tape heads, then an SMM can be represented by a mapping

$$0 \times \Sigma^{k+1} \to Q \times \Sigma^k \times \{L,R,S\}^{k+1} \times K^{2k}.$$

A phase is a period during which no work cell is entered more than once by the heads of the finite control. The space is the total number of work cells used.

Prove that the model SMM (reversal, space) is similar to other models.

Part three Computational types and duality
8 Logical computational types

§35 NON-DETERMINISTIC TURING MACHINES AND NON-DETERMINISTIC VECTOR MACHINES

For simplicity, from now on we treat the machines of all kinds of computational models as language accepters (except a few specified cases).

The Turing machines discussed in Chapter 4 are "deterministic", i.e., starting from any ID, there is at most one next ID. The action of the machine is completely determined. In other words, the next move function δ is a single valued mapping

$$\delta : Q \times \Sigma^k \to Q \times \Sigma^k \times \{L,R,S\}^{k+1}.$$

But if the function δ is a multi-valued mapping from $Q \times \Sigma^k$ to $Q \times \Sigma^k \times \{L,R,S\}^{k+1}$, or equivalently, a single valued mapping from $Q \times \Sigma^k$ to a subset of $Q \times \Sigma^k \times \{L,R,S\}^{k+1}$:

$$\delta : Q \times \Sigma^k \to 2^{Q \times \Sigma^k \times \{L,R,S\}^{k+1}}$$

then, starting from an ID, the machine may have several different next ID's. In other words, the next movement of the machine is not completely determined. There may be (finitely) many different choices. For a fixed input w there are two possibilities:

1. There exists a sequence of choices by which the machine enters an accepting state.

2. For any sequence of choices, the machine does not enter any accepting state.

When case 1 happens, we say the machine accepts the input w, otherwise the machine rejects the input w. A Turing machine obtained this way is called non-deterministic. We shall denote the non-deterministic Turing machine by NTM and denote the deterministic Turing machine by DTM.

Suppose that M is an NTM. We define the language accepted by M, denoted by L(M), to be the set of all input words with which there is some sequence of choices that causes the machine M to arrive at an accepting state. In order to discuss the relation between TM and other models, we always use the same input alphabet {0,1} as before.

DEFINITION 35.1 Suppose that M is an NTM. For a fixed input word $w \in \{0,1\}^*$ and any sequence of choices, let t(w) be the maximum number of steps M takes for w from the beginning to the end of the computation; let s(w) be the maximum number of squares M uses on its work tapes for input w from the beginning to the end of the computation; let r(w) be the maximum number of reversals the work tape heads pass for input w from the beginning to the end of the computation.

Furthermore, define

$$t(n) = \text{Max}\{t(w) \mid |w| \leq n\}$$

$$s(n) = \text{Max}\{s(w) \mid |w| \leq n\}$$

$$r(n) = \text{Max}\{r(w) \mid |w| \leq n\}$$

to be the time complexity, the space complexity and the reversal complexity respectively.

Notice that even when a word w is accepted by an NTM, t(w) may still be infinite, by our definition. If $t(n) \neq \infty$, then for any choice sequence, the machine either accepts within t(n) steps, or halts and rejects within t(n) steps. According to our definition the complexities are in fact the worst case complexities.

Obviously, DTM is a special case of NTM, therefore every language accepted by a DTM can be accepted by an NTM with the same time, space and reversal complexities.

THEOREM 35.1 The languages accepted by NTM's are recursively enumerable.

Proof: suppose that M is an NTM accepting L. Without loss of generality, we can assume that in each step the NTM has at most c different choices. Use i ($0 \leq i \leq c-1$) to represent the i-th choice. We call $i_1 i_2 \ldots i_t$, ($0 \leq i_j \leq c-1$) a choice sequence, which represents that the j-th movement of the machine is determined according to the i_j-th choice: if there is no i_j-th choice at the j-th movement, the machine rejects the input for the given choice sequence.

Now we can design a DTM to simulate M as follows:

```
BEGIN
    t := 1;
 *:notyet := 0;
   FOR all choice sequences C of length t DO
   (simulate M on input w for C;
      IF M accepts w THEN return 1;
      IF M has not yet rejected w THEN notyet := 1)
   IF notyet = 0 THEN return 0 ELSE t := t+1;
   GOTO*
END.
```

If M accepts w then this DTM accepts w, else it rejects (may or may not halt);

THEOREM 35.2 If a language L can be accepted by an NTM of time complexity $t(n) \neq \infty$, then it can be accepted by a DTM in time $O(c^{t(n)})$, where c is a constant. Therefore the language is recursive.

Proof: in fact the DTM in the proof of the preceding theorem will halt within time $O(c^{t(n)})$ for some constant c, because the NTM will halt in time $t(n)$ for any choice sequence.

THEOREM 35.3 (Savitch). Suppose that M is an NTM with space complexity $O(g(n))$, where $g(n)$ satisfies

(1) $g(n) \geq \log n$
(2) the unary representation of $g(n)$ can be computed from 1^n by a DTM in space $O(g^2(n))$.

Then there is a DTM of space $O(g^2(n))$, which can simulate M.

Proof: without loss of generality, we can assume that M has a unique accepting state q and when M enters q, its input head points to the leftmost symbol of the input and its work tapes are all blanks. Thus, for any input word w, there is only one accepting configuration D_a.

A configuration is called an s-configuration, if its work space used is not more than s. We use the following recursive procedure to test whether or not there is a path, which passes through only s-configurations, connecting two s-configurations D_1 and D_2, and is of length $\leq 2^t$; in other words, to test whether or not we can arrive at D_2 from D_1 within space s and time 2^t.

    FUNCTION CONNECT (D_1, D_2, s, t)

    BEGIN

        IF t=0 and $(D_1 = D_2$ or $D_1 \vdash D_2)$ THEN return 1;

        IF t > 0 THEN

            FOR all s-configuration D DO

                IF CONNECT $(D_1, D, s, t-1) = 1$ and CONNECT $(D, D_2, s, t-1) = 1$ THEN return 1;

        return 0;

    END.

There are at most $2^{cg(n)}$ g(n)-configurations. The whole algorithm of the simulating DTM is as follows:

    BEGIN

        D_0 := initial configuration;

        D_a := accepting configuration;

        s := g(n);

        t := cg(n);

        CONNECT (D_0, D_a, s, t)

    END

To compute g(n), we need space $O(g^2(n))$. The depth of the recursive function CONNECT is $O(g(n))$. At each level of recursive calls, we need $O(g(n))$ space to store two configurations. Therefore the whole space used is $O(g^2(n))$.

THEOREM 35.4 For any recursively enumerable language L, there is an NTM of reversal 2 to accept it.

Proof: we can assume that L is accepted by a single-tape TM M, whose ID is a word $\alpha q \beta$, where $\alpha, \beta \in \Sigma^*$, $q \in Q$. An effective computation on input w of M is defined to be

$$D_0 \# D_1 \# \ldots \# D_t$$

where $D_0 = q_0 w$ is the initial ID, D_t is an accepting ID, and $D_i \vdash D_{i+1}$ ($i = 0, 1, \ldots, t-1$). Construct an NTM M' as follows: M' has two work tapes. For input w, in one phase it non-deterministically guesses the same word $\gamma = D_0 \# D_1 \# D_2 \# \ldots \# D_t$ ($D_i \in \Sigma^* Q \Sigma^*$, D_0 is the initial ID for input w) on its two work tapes. Then in another phase (from right to left) it checks whether or not γ is an effective computation of M. (This can be done if we let the first tape head go one ID ahead of the second tape head.) If γ is an effective computation then M' accepts w. Machine M accepts w iff there is an effective computation. The later holds iff M' accepts w in two reversal.

Notice that the time complexity is infinite under our definition.

In the following we discuss non-deterministic vector machines. A non-deterministic vector machine (NVM) differs from deterministic vector machines by allowing that if the fan-out of a vertex in its program (a directed graph) is two then the two edges out from it may be labelled with two opposite predicates as well as two different instructions (for a DVM, if the fan-out of a vertex is two then they can only be labelled with two opposite predicates) and that all vertices of fan-out 0 are partitioned into accepting states and rejecting states.

The NVM acts as follows. It starts from the starting state with the initial ID and goes along any path in the program, until it meets a vertex of fan-out 0. Whenever it passes an edge, if an instruction is labelled on the edge, then it must execute the instruction on the ID; if a predicate is labelled on the edge, then the predicate must be satisfied by the ID. If there is a path going from the starting state to an accepting state, then the NVM accepts the input w. Otherwise the machine rejects.

The maximum number of steps the machine takes for any input word w of length $\leq n$ and any possible path, is the time complexity. The maximum number of the total used register length for any input word w of length $\leq n$ and any possible path, is the space complexity.

Therefore in each step of the computation of the NVM, there may be more than one choice. If the choice sequences are different, then the results of the computation may be different. Some result in accepting, some result in rejecting. The machine accepts iff at least one path results in accepting.

It is easy to prove that the language accepted by NVM is recursively enumerable. But we should notice that the length of the choice sequence of the NTM amounts to the sequential time whereas the length of the choice sequence of the NVM amounts to the parallel time, which is shorter than the sequential time in many cases. This makes a big difference. We have already seen that to simulate an NTM, the DTM needs exponential time. But we shall see that to simulate an NVM, the DVM needs only polynomially related time.

Exercises

35.1 In many papers, the non-deterministic time for NTM is defined as follows.
Let w be the input word of length n. Let $t(w)$ be the smallest number of steps for any choice sequence leading to an accepting state. If there is no such choice sequence, then let $t(w) = 0$. Define $t(n) =$ Max $\{t(w) \mid |w| \leq n\}$ to be the time complexity of the NTM.
Suppose that $1^{f(n)}$ can be computed from 1^n by a DTM within time $O(f(n))$.
Prove that the language classes accepted by NTM's within $O(f(n))$ time under these two different definitions are the same. Prove the similar result for space complexity.

35.2 Prove that the language accepted by an NTM with time complexity $t(n) < \infty$ is recursive.

35.3 (1) Prove that the language accepted by an NTM with space complexity $s(n) < \infty$ can be accepted by an NTM that has the same space complexity and halts for every input. (Therefore the language is recursive.)
(2) Is there a similar result for reversal?

35.4 Prove that every NTM M can be simulated by an NTM M' satisfying
(1) The next move function δ of M' always has 0 or two values, i.e. if the machine M' does not halt, it has exactly two choices.
(2) The time, space, reversal complexities of M' are not more than a constant × the corresponding complexities of M.

35.5* Prove that a language accepted by an NTM in space $s(n)$ and time $t(n)$ can be accepted by a DTM in space $O(s(n)\log(t(n)/s(n)))$.

35.6 Prove that a language accepted by a DTM of time complexity $O(2^n)$ can be accepted by an NTM within $O(n)$ reversal and $O(4^n)$ space.

§36 THE LOGICAL COMPUTATIONAL TYPES FOR TM

We have introduced NTM. Now we generalize the non-deterministic type to logical computational types.

For simplicity, assume that M is an NTM whose every action has two possible choices (if the machine does not halt), i.e., for every ID D there are two next ID's, D' and D". We can define "good" ID's recursively as follows:

(1) accepting ID's are good;
(2) if $D \vdash D'$, $D \vdash D"$, D' is good or D" is good, then so is D;

Then, M accepts input w iff the initial ID for w is good.

In the above definition, whether or not D is good depends on the "or" operation.

$\quad$ D is good $\leftrightarrow$ D' is good or D" is good.

If we change the "or" operation to "and" operation and define

$\quad$ D is good $\leftrightarrow$ D' is good and D" is good,

then we get a new type TM — the co-nondeterministic type TM, or conjunction TM. We can define equivalence TM, exclusive or TM, and so on. The non-deterministic TM will be called disjunction TM also.

To find all logical computational types, we list all sixteen Boolean functions (or operations) of two variables.

τ $(a\tau b = 1)$ $\qquad$ $\bar{\tau}$ $(a\bar{\tau}b = 0)$

$\pi_1 (a\pi_1 b = a)$ $\qquad$ $\bar{\pi}_1 (a\bar{\pi}_1 b = \neg a)$

$\pi_2 (a\pi_2 b = b)$ $\qquad$ $\bar{\pi}_2 (a\bar{\pi}_2 b = \neg b)$

$\vee$ $(a \vee b = 0$ if $a=0$ and $b=0$ $\qquad$ $\triangledown$ $(a \triangledown b = \neg(a \vee b))$

$\quad a \vee b = 1$ else)

$\wedge$ $(a \wedge b = \neg(\neg a \vee \neg b))$ $\qquad$ $\triangle$ $(a \triangle b = \neg(a \wedge b))$

$\rightarrow$ $(a \rightarrow b = \neg a \vee b)$ $\qquad$ $\leftrightarrow$ $(a \leftrightarrow b = \neg(a \rightarrow b))$

$\leftarrow$ $(a \leftarrow b = b \rightarrow a)$ $\quad\quad$ $\nrightarrow$ $(a \nrightarrow b = \neg(b \rightarrow a))$

$\leftrightarrow$ $(a \leftrightarrow b = (a \rightarrow b) \wedge (b \rightarrow a))$ $\quad\quad$ θ $(a \theta b = \neg(a \leftrightarrow b))$

REMARK 36.1 Sometimes we can extend the range of Boolean variables to have three values, 1, 0, or ⌐(undefined). Suppose that * is any Boolean operation mentioned above. If both variables a,b are defined, then so is a*b. But when a or b is not defined, the value of a*b can still be defined. For example, we can define 1v ⌐ = 1, because no matter what b is, we always have 1vb = 1. Generally, if no matter the undefined variable(s) equal(s) to 0 or 1, the result of the operation is always the same, then it is defined to be this value. We shall use this definition later.

DEFINITION 36.1 A logical computational type T is a non-empty subset of the set of the above sixteen Boolean functions.

Therefore there are $2^{16} - 1 = 65535$ logical types.

DEFINITION 36.2 A k work tape T type TM is an 11-tuple.

$$M = (Q, I, \Sigma, \sqcup, \delta_0, \delta_1, q_0, F, G, \phi, T)$$

where

$\quad$ Q is the finite state set;

$\quad$ $q_0 \in Q$, initial state;

$\quad$ $F \subseteq Q$, accepting set;

$\quad$ $G \subseteq Q$, rejecting set; G has no common element with F;

$\quad$ Σ, a finite set, the work tape alphabet;

$\quad$ $I = \{0,1\} \subseteq \Sigma$, input alphabet;

$\quad$ $\sqcup \in \Sigma - I$, blank symbol;

$\quad$ δ_0, δ_1: two mappings: $Q \times \Sigma^{k+1} \rightarrow Q \times \Sigma^k \times \{L,R,S\}^{k+1}$, the next move functions

$\quad$ T, a logical computational type;

$\quad$ ϕ, a mapping $Q \rightarrow T$, for each $q \in Q$, $\phi(q)$ is the type of the state q.

We use different states to determine whether the machine accepts the input, therefore we do not need an output tape. Whenever the machine enters

an accepting state or a rejecting state, it halts. We shall further assume
that the input tape head of the machine will not leave the input word and the
two blank squares adjacent to it.

DEFINITION 36.3 The computation tree of a T type TM for input w is a (possibly infinite) binary tree, whose nodes have the form (x, D), where
$x \in \{0,1\}^*$, D is a configuration such that

(1) the root of the tree is (Λ, D_0), where D_0 is the initial configuration for w;

(2) if (x, D) is a node of the computation tree, D is an accepting configuration or a rejecting configuration, then (x,D) has no son;

(3) if (x,D) is a node of the computation tree, and D is neither an accepting configuration nor a rejecting configuration, then (x,D) has a left-hand son (x0,D') and right-hand son (x1,D"), where D' is the next configuration of D by the next move function δ_0, D" is the next configuration of D by the next move function δ_1.

In other words, the x in node (x,D) is the choice sequence mentioned in the last section, while D is the configuration obtained from the initial ID according to the choice sequence x. The node (x,D) is completely determined by x, therefore we also denote node (x,D) by x. Sometimes we use x to denote the path from the root to node (x,D).

From now on, the state of node (x,D) means the state q in D. The type of node (x, D) means the type $\phi(q)$ of q.

Obviously, node (x,D) is a leaf of the computation tree iff D is an accepting or rejecting configuration.

Since the computation tree is possibly infinite, we give every node a (0,1)-assignment ψ (a mapping from some nodes to $\{0,1\}$), as follows.

1. For every natural number m, define ϕ_m for each node in level m, level m-1,..., level 1, level 0 as follows:

If node x is in level m, then

$$\psi_m(x) = \begin{cases} 1, & \text{if the state of x is accepting,} \\ 0, & \text{if the state of x is rejecting,} \\ \text{undefined,} & \text{otherwise} \end{cases}$$

Suppose that $\psi_m(x)$ is defined for nodes in level i+1, we define the value of $\psi_m(x)$ for node x in level i by

$$\psi_m(x) = \begin{cases} 1, & \text{if the state of x is accepting,} \\ 0, & \text{if the state of x is rejecting,} \\ \psi_m(y_1)\phi(q)\psi_m(y_2), & \text{where } y_1 \text{ and } y_2 \text{ are the left-hand and} \\ & \text{right-hand sons of node x, } \phi(q) \text{ is the type of node x.} \end{cases}$$

The operation $\phi(q)$ should be explained according to Remark 36.1.

2. Obviously, ψ_{m+1} is an extension of ψ_m, i.e., if a node is defined for ψ_m, then it is defined for ψ_{m+1}, and has the same value as ψ_m. Now define $\psi = \lim \psi_m$ as follows:

$\psi(x) = a$, if there exists m such that $\psi_m(x) = a$,

$\psi(x) =$ undefined, otherwise.

It is easy to see that $\psi(x)$ is well defined.

DEFINITION 36.4 ψ is called the minimum assignment of the computation tree. If the value $\psi(r)$ of the root r of the assignment is 1, then the machine accepts the input. Otherwise M rejects.

ψ is called the minimum assignment because it has the smallest defining field among all assignments satisfying the condition that the value at any node must be the result of its operation on values of its two sons.

THEOREM 36.1 Let T be any logical computational type. Then the language accepted by a TM of type T is recursively enumerable.

Proof: in fact, in the definition of the minimum assignment ψ, we have given a method to simulate a TM of type T by a DTM: for any m, construct a computation tree of height m, and compute the value ψ_m at the root. If there is an m such that the value of ψ_m on the root is 1, then the input is accepted.

REMARK 36.2 The Boolean functions τ, $\bar{\tau}$, π_1, π_2, $\bar{\pi}_1$, $\bar{\pi}_2$ are in fact "deterministic", because their values are determined by only one argument. Therefore if a state is in one of these types, then only one selection is meaningful. All the other states are non-deterministic.

DEFINITION 36.5 Suppose that for w the maximum number of non-deterministic states that the computation has passed along any path is c(w). The function

$$c(n) = \text{Max}\{c(w) \mid |w| \leq n\}$$

is called the choice complexity. If $h(n) \geq c(n)$ is a log-space constructible function, then we say the machine is of choice complexity $h(n)$.

Exercises

36.1 If $T = \{\tau\}$, $\{\bar{\tau}\}$ or $\{\tau,\bar{\tau}\}$, what is the language accepted by TM of type T?

36.2 Suppose $T = \{\pi_1\}$ or $\{\pi_2\}$, what is the relation between TM of type T and deterministic TM?

36.3 Suppose $T = \{\bar{\pi}_1\}$ or $\{\bar{\pi}_2\}$, what is the relation between TM of type T and deterministic TM?

§37 THE CLASSIFICATION OF LOGICAL COMPUTATION TYPES

We can see from the preceding section that it is not meaningful to introduce logical computational types when we want to discuss only theoretical computability. From now on, we shall therefore consider the relationships between different computational types according to their complexities.

DEFINITION 37.1 Suppose that M is a Turing machine of type T. The maximum height of computation trees for all input words of length $\leq n$ is the time complexity $t(n)$; the maximum number of work tape squares the machine uses, along any path starting at the root for all input words of length $\leq n$, is the space complexity $s(n)$; the maximum number of reversals the work tape heads pass, along any path starting at the root for all input words of length $\leq n$, is the reversal complexity $r(n)$.

DEFINITION 37.2 Suppose that T_1 and T_2 are two logical computational types. If for any TM M_2 of type T_2, there is a TM M_1 of type T_1 simulating T_2 such that

$$r_1(n) = O(r_2(n)), \quad s_1(n) = O(s_2(n)), \quad t_1(n) = O(t_2(n)),$$

where r_1, s_1, t_1 are the reversal, space and time complexities for M_1; r_2, s_2, t_2

are the reversal, space and time complexities for M_2: then we say that T_1 can linearly simulate T_2, $T_1 \overset{L}{\geq} T_2$.

If T_1 and T_2 can linearly simulate each other, then we say that T_1 and T_2 are linearly similar, denoted by $T_1 \overset{L}{\cong} T_2$.

In the following we discuss only the case when $t(n) < \infty$, i.e., for any input w, the computation tree is finite. Therefore we have $s(n) < \infty$ and $r(n) < \infty$.

When $T = \{\tau\}$, $\{\bar{\tau}\}$ or $\{\tau\ \bar{\tau}\}$, the language accepted by a TM of type T is the full set $(0+1)^*$ or empty set $\emptyset$. Therefore these three types are trivial. The other 25532 types are non-trivial.

DEFINITION 37.3 Suppose that T_1 and T_2 are two logical computational types. If any function in T_2 can be obtained by variable substitution and function composition from the functions in $T_1 \cup \{\tau,\bar{\tau}\}$, then we denote as $T_1 \geq T_2$. If $T_1 \geq T_2$ and $T_2 \geq T_1$, then we say they are equivalent, denoted by $T_1 \cong T_2$.

For example, $\pi_1(a,b) = a = \pi_2(b,a)$, therefore $\{\pi_1\} \cong \{\pi_2\}$. $\wedge(a,b) = \neg(\neg a \vee \neg b) = \bar{\pi}_1(\vee(\bar{\pi}_1(a,a), \bar{\pi}_1(b,b)),a)$, therefore $\{\wedge\} \leq \{\bar{\pi}_1, \vee\}$. Since $\neg a = \rightarrow (a,0) = \rightarrow (a,\bar{\tau}(a,a))$, we have $\{\bar{\pi}_1\} \leq \{\rightarrow\}$. Therefore $\{\vee\} \leq \{\rightarrow\}$ and $\{\vee\} \cong \{\rightarrow\}$.

LEMMA 37.1 If $T_1 \geq T_2$ then $T_1 \overset{L}{\geq} T_2$.

Proof: since $T_1 \geq T_2$, each function in T_2 can be obtained by substitution and composition from the functions in T_1 and $\{\tau,\bar{\tau}\}$. A state of type τ or $\bar{\tau}$ can be realized by an accepting state or a rejecting state respectively. A state of a type in T_2 can be simulated by a composition of some constant number of states of types in $T_1 \cup \{\tau,\bar{\tau}\}$. The reversal and the space complexities are the same. The time complexity of the simulating TM of type T_1 is a constant × the time complexity of the TM of type T_2 being simulated.

LEMMA 37.2 If we use $T_1 \rightarrow T_2$ to express $T_1 \overset{L}{\geq} T_2$, then we have the relationships shown in Fig. 37.1 and any non-trivial logical computational type is linearly similar to one type in this figure.

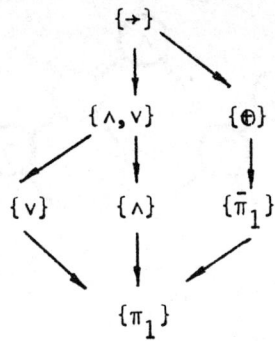

Fig. 37.1

Proof: the relations between logical types displayed in Fig. 37.1 is obvious. To prove that every non-trivial logical computational type is linearly similar to one of the seven types, we need only to show that adding any operation to any one of the seven types will still result in a type equivalent to one of the seven types. The details are left as an exercise.

LEMMA 37.3 $\{\rightarrow\} \stackrel{L}{=} \{\wedge,\vee\}$

Proof: since $\{\nabla\} \cong \{\rightarrow\} \geq \{\wedge,\vee\}$, we need only to prove that $\{\nabla\} \stackrel{L}{\leq} \{\wedge,\vee\}$.

The idea behind the proof is that in a Boolean expression formed by some $\neg, \wedge, \vee$ operations, we can always remove the $\neg$ operation everywhere except directly ahead of a variable, by the following formulas

$$\neg(a \vee b) = \neg a \wedge \neg b$$
$$(a \wedge b) = \neg a \vee \neg b$$
$$\neg \neg a = a$$

Therefore the following "nor tree" can be transferred to a $\wedge$, $\vee$ tree step by step, as shown in Fig. 37.2.

171

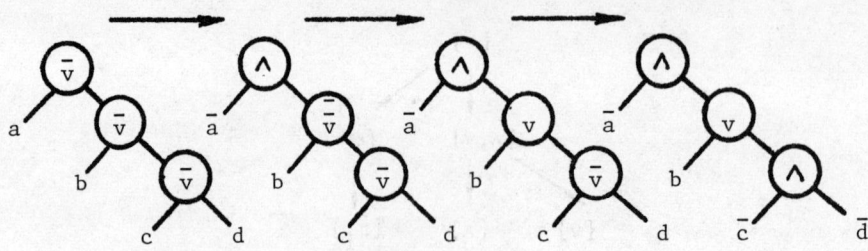

Fig. 37.2

Formally, suppose that M is a TM of type $\{\vee\}$. Let Q be its state set, δ_0, δ_1 the next move functions, q_0 the initial state, and F, G the accepting state set and the rejecting state set respectively.

Define TM M' of type $\{\wedge, \vee\}$ as follows:

state set $Q' = Q \times \{0,1\} = \{[q,0], [q,1] \mid q \in Q\}$

initial state $[q_0, 0]$;

accepting state set $F' = F \times \{0\} \cup G \times \{1\}$;

rejecting state set $G' = G \times \{0\} \cup F \times \{1\}$.

If the next move functions of M are

$$\delta_0(q, a_1, \ldots, a_{k+1}) = (p, b_1, \ldots, b_k, d_1, \ldots, d_{k+1});$$
$$\delta_1(q, a_1, \ldots, a_{k+1}) = (\bar{p}, \bar{b}_1, \ldots, \bar{b}_k, \bar{d}_1, \ldots, \bar{d}_{k+1}),$$

then define

$$\delta_0'([q,0], a_1, \ldots, a_{k+1}) = ([p,1], b_1, \ldots, b_k, d_1, \ldots, d_{k+1})$$
$$\delta_0'([q,1], a_1, \ldots, a_{k+1}) = ([p,0], b_1, \ldots, b_k, d_1, \ldots, d_{k+1})$$
$$\delta_1'([q,0], a_1, \ldots, a_{k+1}) = ([\bar{p}, 1], \bar{b}_1, \ldots, \bar{b}_k, \bar{d}_1, \ldots, \bar{d}_{k+1})$$
$$\delta_1'([q,1], a_1, \ldots, a_{k+1}) = ([\bar{p}, 0], \bar{b}_1, \ldots, \bar{b}_k, \bar{d}_1, \ldots, \bar{d}_{k+1})$$

All the states $[q,0]$ are of type $\wedge$. All the states $[q,1]$ are of type $\vee$.

Thus, the action of M' is about the same as M. They are different only in that the states of M' will turn from type ∧ to type ∨, and from type ∨ to type ∧ alternately.

When a node x is at an even level, if the state of x in the computation tree of M is q, then the state of node x in the computation tree of M' is [q,0]. But when x is at an odd level, the state of x in the tree of M' is [q,1]. It is not difficult to prove recursively that (as an exercise)

(1) at even levels, the corresponding nodes of these two trees have the same assignment;

(2) at odd levels, the corresponding nodes of these two trees have different assignments.

Therefore the two roots of these two trees have the same assignment. Hence they accept the same language and use the same reversal, space and time.

Notice that the machine M' has the following properties.

1. The initial state is of type ∧.
2. If the machine is in a state of type ∧, then next time it will enter a state of type ∨. If the machine is in a state of type ∨ then next time it will enter a state of type ∧.

We call this kind of TM a standard {∧,∨} TM.

COROLLARY 37.1 A TM of type {∧,∨} can be linearly simulated by a standard {∧,∨} TM.

LEMMA 37.4 $\{\pi_1\} \stackrel{L}{=} \{\bar{\pi}_1\}$.

Proof: since $\{\pi_1\} \stackrel{L}{\leq} \{\bar{\pi}_1\}$, we need only to prove that $\{\bar{\pi}_1\} \stackrel{L}{\leq} \{\pi_1\}$. The proof is similar to that of Lemma 37.3 and is left as an exercise (Exercise 36.3).

THEOREM 37.1 If we use $T_1 \rightarrow T_2$ to express $T_1 \stackrel{L}{\geq} T_2$, then we have the relationships shown in Fig. 37.3 and any non-trivial logical computational type is linearly similar to one of the five types shown.

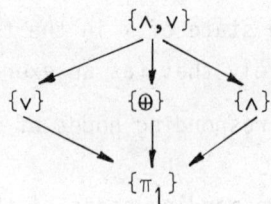

Fig. 37.3

The type $\{\wedge,\vee\}$ is called alternating. The type $\{\pi_1\}$ is in fact deterministic. The type $\{\theta\}$ is called equivalent, because it is linearly similar to type $\{\leftrightarrow\}$. We shall denote TM of type $\{\wedge,\vee\}$, $\{\vee\}$, $\{\wedge\}$ $\{\theta\}$, $\{\pi_1\}$ by ATM, NTM, CTM, ETM and DTM respectively.

Exercises

37.1 If a language L can be accepted by an NTM within r(n) reversal, s(n) space and t(n) time, then the complement $(0+1)^*$-L can be accepted by a CTM within the same reversal, space and time, and vice versa.

37.2 Prove the same result for type $\{\theta\}$ and type $\{\leftrightarrow\}$. But we have $\{\theta\} \stackrel{L}{=} \{\leftrightarrow\}$. What conclusion can we obtain?

37.3 Given a coding of an undirected graph of n nodes and a binary representation of an integer k, prove that the following problems can be computed by an NTM, CTM and ETM in polynomial time respectively.

  (1) Is there a k-complete subgraph?
  (2) Is there no k-complete subgraph?
  (3) Is the total number of k-complete subgraphs even?

37.4 Prove Lemma 37.2.

37.5 In Lemma 37.1, if $\{\bar{\pi}_1,\vee\} \subseteq T_1$ and $\wedge \in T_2$, then how can you simulate one state of type $\wedge$ by some states of type $\bar{\pi}_1$ and type $\vee$? Can you use not more than four states?

§38 ALTERNATING TURING MACHINES

For ATM's introduced in the last section, each configuration has exactly two conjuctive choices or two disjunctive choices. We can generalize the definition to allow each configuration to have finitely many conjunctive choices or finitely many disjunctive choices. The generalized alternating type TM is still linearly similar to the standard alternating type TM. The following four theorems are due to Chandra and Stockmeyer.

THEOREM 38.1 If a language L can be accepted by an ATM in time $t(n) \geq n$, then it can be accepted by a DTM in $O(t(n))$ space.

Proof: we do depth-first-search on the computation tree of the ATM.

Suppose that the language L is accepted by an ATM M in time $t(n)$. Let STATE(x,w) be a subprogram which simulates M and outputs the state that M will arrive given input w and choice sequence x. Obviously the space used is $O(t(n))$. Then use the following recursive program to determine the assignment of node x in the computation tree for input w:

```
FUNCTION  ACCEPT(w,x)
BEGIN
   q := STATE(w,x);
   IF q is accepting THEN return 1;
   IF q is rejecting THEN return 0;
   IF q is disjunctive THEN return ∨ ACCEPT(w,xi)
                                    i
                       ELSE return ∧ ACCEPT(w,xi)
                                    i
END.
```

Then ACCEPT $(w,\wedge)$ is the program of the simulating DTM. The depth of the recursive calls is $O(t(n))$. At each level of the recursive calls we need $O(1)$ space (except the space needed for subprogram STATE, which needs $O(t(n))$ space altogether). Therefore the total space used is $O(t(n))$.

THEOREM 38.2 Suppose that M is a DTM accepting L with space complexity $s(n) \leq g(n)$, where $g(n) \geq n$ is log-space constructible. Then there is an ATM of time completely $O(g^*(n))$ to accept L.

Proof: let c be a positive integer satisfying that the time used by M is not more than $2^{cg(n)}$ and the length of configurations of M is not more than $cg(n)$.

Construct an ATM M' as follows: first, M' constructs three intervals J_1, J_2 and J_3; each is of length $cg(n)$ and is used to store a configuration of M. The above work needs time $O(g^*(n))$, since $g(n) \geq n$ is log-space constructible.

Secondly, M' puts the initial ID of M into J_1, enters a disjunctive state, selects an accepting configuration and puts it into J_3. This needs time $O(g(n))$.

Finally, M' performs the following four steps for $cg(n)$ iterations and then goes into a rejecting state.

1. Enters a disjunctive state, jumps to 2 or 3.
2. If the configuration in J_3 is the next configuration to that in J_1, then accepts, else rejects.
3. Enters a disjunctive state, selects a configuration of length $\leq cg(n)$ and puts into J_2.
4. Enters a conjunctive state to do the following two tasks:
   (i) exchange the contents of I_3 and I_2;
   (ii) exchange the contents of I_1 and I_2.

If M accepts w, then there must be a computation

$$I_0 \vdash I_1 \vdash I_2 \vdash \ldots \vdash I_t,$$

where I_0 is the initial configuration and I_t is an accepting configuration, $t \leq 2^{cg(n)}$. Therefore there must be a configuration $I_{[t/2]}$ "in the middle" such that $I_1 \vdash I_{[t/2]}$ and $I_{[t/2]} \vdash I_t$. Step 3 is to guess such a configuration $I_{[t/2]}$. Step 4 is to verify both $I \vdash I_{[t/2]}$ and $I_{[t/2]} \vdash I_t$, recursively. This technique is often used and is called "guess in the middle and verify both ends".

Obviously the time used by the ATM M' is $O(g^*(n))$. Notice that if we could compute $g(n)$ faster, then the whole time used by M' might be only $O(g^2(n))$.

THEOREM 38.3 If L can be accepted by an ATM in space $s(n)$, then it can be accepted by a DTM in time $O(n^* c^{s(n)})$, for some constant $c > 1$.

Proof: suppose that M is a standard ATM accepting L with space complexity $s(n)$. We view the computation of M as a directed graph G, whose nodes are configurations. The fan-out of each node is two or zero. If configuration D' can be obtained by one step of the computation from configuration D, then there is an arrow from node D to node D' in G (whether the type of D be ∧ or ∨).

We design a DTM M' as follows. First, M' constructs the initial configuration D_0 of M. Then it constructs all direct successors of D_0, all direct successors of direct successors of D_0,..., and so on. Since there are only $O(nc_1^{s(n)})$ nodes in G, the simulating DTM can construct the coding of the graph G in time $O(n^*c_2^{s(n)})$. Because there is no cycle in G (otherwise, M will not hold), the DTM M' can give each node a {0,1}-label as follows.

1. If the fan-out of a node is 0, then label it with 1 or 0 according to whether the state of the configuration is accepting or rejecting, respectively.

2. If the fan-out of a node is 2, and the two direct successors have been already labelled, then label this node according to the type of the state of the node and the labels on its two successors.

The DTM M' keeps on doing the above work until no more nodes can be labelled. The total time used by M' is $O(n^*c_3^{s(n)}))$. Finally, M' accepts or rejects the input according to whether the node D_0 is labelled with 1.

In the above proof, the key point is that we construct the computation graph instead of the computation tree. The size of the computation tree could be doubly exponentially large.

THEOREM 38.4 Suppose that M is a DTM accepting L with time complexity $t(n) \leq h(n)$, where $h(n) \geq n$ is log-space constructible. Then there is an ATM M' to accept L within space $O(\log(h(n)))$.

Proof: without loss of generality, we can assume that M has only one tape, one accepting state q_1, and when the machine enters the accepting state q_1 the contents of the tape are all blanks and the tape head is in its initial position.

In the following, we define an ID of M to be a word in $\Sigma^*(Q \times \Sigma)\Sigma^*$ of length $2h(n) + 1$, i.e.,

$$ID = b_{-h(n)}b_{-h(n)+1} \cdots [b_i,q] \cdots b_{h(n)-1}b_{h(n)}$$

representing that the contents of the tape are

$$b_{-h(n)}b_{-h(n)+1} \cdots b_i \cdots b_{h(n)-1}b_{h(n)},$$

the tape head is scanning b_i, and the state is q. For input w, the initial ID is

$$D = \underbrace{\sqcup\sqcup\sqcup\cdots\sqcup\sqcup\sqcup\sqcup [a_1,q]a_2}_{h(n)} \cdots a_n \underbrace{\sqcup\sqcup\sqcup\sqcup \cdots \sqcup\sqcup\sqcup\sqcup\sqcup}_{h(n)-n+1}$$

The computation of M is a sequence of ID's:

$$D_0 \vdash D_1 \vdash D_2 \vdash \cdots \vdash D_{h(n)},$$

where if D_i is an accepting ID then $D_i = D_{i+1} = \cdots = D_{h(n)}$. This sequence has the following property. The j-th symbol ($-h(n) \leq j \leq h(n)$) of D_i is completely determined by the j-1-th, j-th, and j+1-th symbols of D_{i-1}. The symbols may belong to Σ or $Q \times \Sigma$.

For every symbol σ, let $F(\sigma)$ be the set of all sequences $\sigma_{-1}\sigma_0\sigma_1$ such that, if the j-1, j, j+1-th symbol of some ID of M at time t is $\sigma_{-1},\sigma_0,\sigma_1$ respectively, then the j-th symbol of the ID of M at time t+1 will be σ. For each σ, $F(\sigma)$ is a finite set.

We are going to design an ATM M' which can determine whether or not the o-th symbol of $D_{h(n)}$ is $[\sqcup,q_1]$. If it is, then M' accepts; otherwise, M' rejects. Therefore M' can simulate M.

We use the predicate $H(j,t,\sigma)$ to mean that the j-th symbol of D_t is σ. Then (see Figure 38.1)

$H(j, t+1, \sigma)$

$\leftrightarrow$ There exists $\sigma_{-1}\sigma_0\sigma_1 \in F(\sigma)$ such that

$H(j-1, t, \sigma_{-1}) \wedge H(j, t, \sigma_0) \wedge H(j+1, t, \sigma_1)$.

$\leftrightarrow$ There exist $\tau_{-2}\tau_{-1}\tau_0 \in F(\sigma_{-1})$, $\tau_{-1}\tau_0\tau_1 \in F(\sigma_0)$ and $\tau_0\tau_1\tau_2 \in F(\sigma_1)$ such that

$H(j-2,t-1,\tau_{-2}) \wedge H(j-1,t-1,\tau_{-1}) \wedge H(j,t-1,\tau_0) \wedge H(j+1,t-1,\tau_1) \wedge$

$H(j+2, t-1, \tau_2)$

and so on.

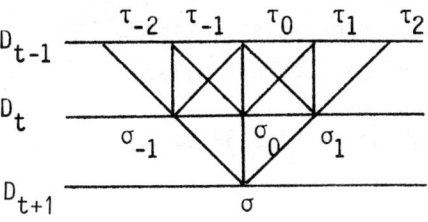

Fig. 38.1

Inspired by this, we can give the program of the ATM as below.

1. $(j,t,\sigma) \leftarrow (0, h(n), [\sqcup, q])$

2. If $t = 0$, then enters an accepting state or rejecting state according to whether the j-th symbol of D_0 is σ. (i.e., whether $H(j,0,\sigma)$ holds).

3. Selects $\sigma_{-1}\sigma_0\sigma_1 \in F(\sigma)$ disjunctively.

4. Selects integer $i = -1, 0, 1$ conjunctively.

5. $(j,t,\sigma) \leftarrow (j+i, t-1, \sigma_i)$, go to 2.

To compute the binary expression of $h(n)$, space $O(\log(h(n)))$ is needed. To store j and t in binary form, $O(\log(h(n)))$ space is enough. If M accepts w, then M' will sure accept w. Conversely, if M' accepts w then we can prove by induction for t that all the symbols correctly selected (for example, $\tau_{-2}, \tau_{-1}, \tau_0, \tau_1, \tau_2$ in Fig. 38.1) are unique and consistent. Therefore M accepts w.

Synthesizing the above four theorems, we see that alternating time corresponds to deterministic space while alternating space corresponds to deterministic exponential time. More precisely, if $g(n) \geq n$ is log-space constructible, then

languages accepted by deterministic space g^*
= languages accepted by alternating time g^*;

languages accepted by deterministic time g^*
= languages accepted by alternating space $O(\log g)$.

If $g(n) \geq \log n$ is constructible in linear space (i.e., there is a DTM to compute $g(n)$ in unary form in space $O(g(n))$), then

languages accepted by deterministic time $2^{O(g)}$
= languages accepted by alternating space $O(g)$

The above results are shown in Fig. 38.2.

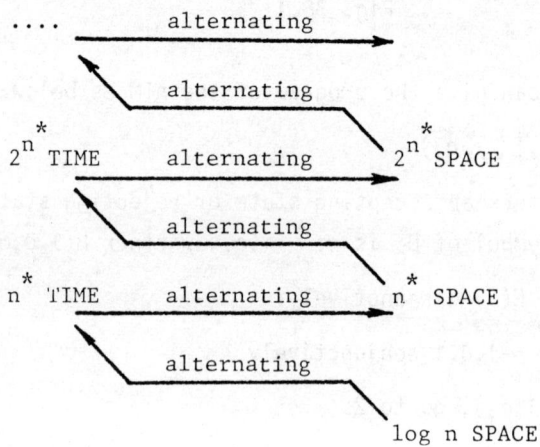

Fig. 38.2

9 General computational types

§39 REFERENCE MACHINE AND THE FIRST SIMILARITY THEOREM

In Chapter 8, we discussed logical computational types. In this chapter we will recharacterize all these concepts at a higher and more general level.

In the discussion of logical computational types in §36, every node of the computation tree of the TM of type T is an ordered pair (x,D), where x is the sequence of choices. D is the instantaneous description word the computation should reach according to the choice sequence x. In other words, for any TM of type T, given choice sequence x and input w, the computation is completely determined by x and w, and the height of the tree is equal to t(w) (the sequential time).

Since we only discuss the worst-case complexities, we propose the following concept of a reference machine.

DEFINITION 39.1 A reference TM is a multitape TM with a read only input tape, a read only reference input tape and k work tapes. At the beginning of the computation, the input word w is on the input tape, and the reference input word x, which is infinite to the right, is on the reference input tape. The pair (w,x) is called the whole input, i.e., the whole input consists of a finite input w and an infinite reference input x. The machine halts for any (w,x) by entering an accepting state or a rejecting state. Therefore, though x is infinite, only a finite part will be scanned.

DEFINITION 39.2 The reversal r(w,x) of M for input (w,x) is the total number of times that the work tape heads change direction (the number of times that the input and/or reference input tape head change direction will not be included). Let INS be the set of all infinite binary sequences. We define

$$r(n) = \text{Max } \{r(w,x) \mid |w| \leq n, x \in \text{INS}\}$$

to be the reversal complexity of M.

The space s(w,x) of M for input (w,x) is the total number of squares used on the work tapes (the number of used squares on the input or reference input tape will not be included). We define

$$s(n) = \text{Max } \{s(w,x) \mid |w| \leq n, x \in \text{INS}\}$$

to be the space complexity of M.

The time $t(w,x)$ of M for input (w,x) is the total number of steps from beginning to the end of the computation. We call

$$t(n) = \text{Max } \{t(w,x) \mid |w| \leq n, x \in \text{INS}\}$$

the time complexity of M.

The choice $c(w,x)$ is the number of the reference input bits being scanned. Define

$$c(n) = \text{Max } \{c(w,x) \mid |w| \leq n, x \in \text{INS}\}$$

to be the choice complexity.

<u>DEFINITION 39.3</u> Suppose that (f,g) is a nice pair, and h is a log-space constructible function. If a reference TM satisfies

$r(n) \leq f(n)$,

$s(n) \leq g(n)$,

$c(n) \leq h(n) \leq f^*(n)g^*(n)$,

then M is an (f,g)-h reference machine.

<u>LEMMA 39.1</u> For any (f,g)-h reference TM, we have $t(n) \leq f^*g^*$.

Proof: by an argument similar to that in Theorem 17.2, we have $t(n) \leq n^*h^*f^*g^* \leq f^*g^*$.

A similar definition of (f,g)-h reference machines for RAM, VM, PRAM can be given (Exercise 39.1). What should be noticed is that for VM, the reference input is stored in a vector V_1 which is not included in the work space. In the narrow case there should be two indirect read devices, one for input, the other for reference input (see Remark 34.1). In the flat case, the reference input can be taken by a ∧ instruction on V_1 and another vector whose contents are $+1^m$ for some integer m. Therefore in either case it is clear which reference bit has been used. For RAM, the total length of index registers is still bounded by $O(\log (fg))$.

For references UC and UA, a few more words are necessary.

Suppose that $h(n)$ is log-space constructible, and (f,g) is a nice pair of functions. An $h(n)$-reference circuit of input length n is a circuit C_n in which some designated n gates are input gates, some other designated $h(n)$ gates are reference input gates. An (f,g)-$h(n)$ reference UC is a family of $h(n)$-reference circuits C_n of input length n satisfying

(1) $h(n) \leq f^*(n)g^*(n)$,

(2) the depth $\leq f(n)$, the width $\leq g(n)$,

(3) the coding of C_n is log-space constructible, given 1^n as input.

The reference UA can be defined similarly and is left as an exercise.

Consider an (f,g)-h reference TM M. Obviously if $|y| = h(|w|)$ then for all $u, v \in INS$, we have $M(w,yu) = M(w,yv)$, because M will never scan u nor v. Therefore we can define $M(w,y) = M(w,yx)$ for any $x \in INS$. Thus for a fixed w, $M(w,y)$, $|y| = h(|w|)$, is a leaf labelled complete binary tree of height $h(n)$.

If a node has two sons with the same label then remove the two sons and put the label on this node. We do the above transformation on the tree $M(w,y)$, $|y| = h(|w|)$, until there is no such a node. Then we obtain a leaf labelled binary tree ϕ_{Mw}.

DEFINITION 39.4 ϕ_{Mw} is called the computation tree of M on w.

Checking the proofs in Chapters 5, 6 and 7, we see that all proofs are suitable for reference machines. Therefore we have

LEMMA 39.2 Suppose that Model and Model' are two reference computational models in the following list. Then for any (f,g)-h reference machine M in Model there is an (f^*,g^*)-h reference machine in Model' such that, for any (w,y) with $|y| = h(|w|)$, we have

$$M(w,y) = M'(w,y)$$

(both machines accept or both reject).

1. TM (reversal, space)
2. RAM (reversal, space)
3. VM (time, space)
4. UC (depth, width)
5. UA (time, space)
6. PRAM (time, space)

DEFINITION 39.5 Suppose that Model and Model' are two computational models. If for any (f,g)-h reference machine M in Model there is an (f^*,g^*)-h reference machine M' in Model' such that for any w they have the same computation tree

$$\phi_{Mw} = \phi_{M'w}$$

then we say that Model can be simultaneously simulated by Model'. If they can simultaneously simulate each other, then we say they are similar.

Notice that in this definition the meaning of "computational model" is not clear. But if we mean the models introduced in this book, the meaning is precise.

THEOREM 39.1 (The First Similarity Theorem). The following computational models of reference machines are similar.

1. TM (reversal, space)
2. RAM (reversal, space)
3. VM (time, space)
4. UC (depth, width)
5. UA (time, space)
6. PRAM (time, space)

Proof: by Lemma 39.2, for any (f,g)-h reference machine in Model there is an (f^*,g^*)-h reference machine in Model' such that M(w,y) = M'(w,y) for |y| = h(n). Therefore $\phi_{Mw} = \phi_{M'w}$.

By Theorem 29.2, Theorem 30.2 (for reference machines) and the First

184

Similarity Theorem, we have

THEOREM 39.2 Every (f,g)-h reference machine can be simulated by an $(g^*, 2^{g^*})$-h reference machine, i.e., for the same input w they have the same computation tree.

THEOREM 39.3 Every (f,g)-h reference machine can be simulated by a $(2^{f^*}, f^*)$-h reference machine, i.e., for the same input w they have the same computation tree.

Exercises

39.1 Give the definitions of (f,g)-h reference machines for RAM, VM, PRAM.

39.2 Suppose that M is a reference TM. Prove that

$t(n) = O(nc(n)r(n)s(n));$

$t(n) = O(nc(n)d^{s(n)})$ for some $d > 1$;

$t(n) = O((dnc(n))^{r(n)+1})$ for some $d > 1$.

39.3 Prove Lemma 39.2, Theorems 39.2, 39.3

39.4 Suppose that (f,g) is an nice pair, h is log-space constructible, and $h \leq f^* g^*$. Prove that there is an (f^*, g^*)-TM to compute the unary representation and the binary representation of $h(n)$, given 1^n as input.

§40 GENERAL COMPUTATIONAL TYPES

In section 36, the computation tree of a Turing machine of type T is a binary tree. We first give every leaf a $\{0,1\}$-assignment, and then according to the function ϕ, by the corresponding logical operation, obtain a $\{0,1\}$-assignment for every node. Thus we obtain a labelled binary tree. The machine accepts or rejects according to whether or not the assignment of the root is 1.

Generalizing the idea, we give the following definitions.

DEFINITION 40.1 Let the set of all finite leaf labelled binary trees be TREE.

A (partial) mapping T from TREE to {0,1} is called a (general) computational type.

DEFINITION 40.2 Suppose that Model is a computational model and T is a computational type. We say that a machine M is in Model of type T if M is an (f,g)-h reference machine in Model for a nice pair (f,g) and a log-space constructible h such that, for any input w, $T(\phi_{Mw})$ is always defined to be 0 or 1, and the machine M under type T accepts w iff $T(\phi_{Mw}) = 1$. In other words, $M(w) = T(\phi_{Mw})$. The set of all input words accepted by M is denoted by L(M).

Notice that $T(\phi_{Mw})$ is always defined but T is a partial mapping. Therefore the binary computation tree ϕ_{Mw} of a type T machine M must be in the domain of T. Now we can give the concept of similarity for computational models under type T:

DEFINITION 40.3 Suppose that Model and Model' are two computational models, T is a computational type. If for any (f,g)-h machine M of type T in Model, there is an (f^*,g^*)-h machine M' of type T in Model' such that for any input w we have

$$M(w) = M'(w)$$

(both machines accept or both reject), then we say that Model of type T can be simultaneously simulated by Model' of type T. If they can simultaneously simulate each other, then we say they are similar.

THEOREM 40.1 (The Second Similarity Theorem). For any computational type T, the six models listed in Theorem 39.1 of type T are similar.

Proof: suppose that M is an (f,g)-h machine of type T in Model. By the First Similarity Theorem, there is an (f^*,g^*)-h reference machine M' in Model' such that, for any w, their leaf labelled computation trees are the same: $\phi_{Mw} = \phi_{M'w}$. They are mapped to the same value (0 or 1) by the same mapping T : $T(\phi_{Mw}) = T(\phi_{Mw})$. Thus

$$M(w) = M'(w).$$

By Definition 40.3, the Model of type T can be simultaneously simulated by Model' of type T.

DEFINITION 40.4 (Type Td). For any leaf labelled binary tree ϕ, if all its leaves are labelled by 1 then $Td(\phi) = 1$, if all its leaves are labelled by 0 then $Td(\phi) = 0$. Otherwise $Td(\phi)$ is undefined.

THEOREM 40.2 Type Td is deterministic. More precisely,

(1) suppose that M is an (f,g)-TM, then for any log-space constructible $h \leq f^*g^*$, there is an (f,g)-h reference TM of type Td accepting L(M);

(2) suppose that M is an (f,g)-h reference TM of type Td, then there is a (deterministic) (f,g)-TM accepting L(M).

The same is true for other models (RAM,VM,UA,UC,PRAM,...).

Proof: (1) In fact M is an (f,g)-0 reference TM

(2) We use a deterministic TM M' to simulate M. M' assumes that all reference input bits of M are 0's. Then M' simulates M and is an (f,g) TM.

DEFINITION 40.5 (Type Tn). For a leaf labelled binary tree $\phi \in$ TREE, define $Tn(\phi) = \bigvee_x \phi(x)$, i.e., $Tn(\phi) = 1$ iff there is a leaf of ϕ, whose label is 1.

DEFINITION 40.6 (Type Tc). For a leaf labelled binary tree $\phi \in$ TREE, define $Tc(\phi) = \bigwedge_x \phi(x)$, i.e., $Tc(\phi) = 1$ iff all leaves are labelled with 1.

DEFINITION 40.7 (Type Te). For a leaf labelled binary tree $\phi \in$ TREE, define $Te(\phi) = \leftrightarrow_x \phi(x)$, i.e., $Te(\phi) = 1$ iff the number of leaves of ϕ whose labels are 0 is even.

DEFINITION 40.8 (Type Ta). For a leaf labelled binary tree $\phi \in$ TREE we assume all nodes at even levels are conjunctive, all nodes at odd levels are disjunctive. We compute the values for all nodes from the leaves to the root, level by level. The value of the root is $Ta(\phi)$.

The relations between Tn, Tc, Te, Ta and the logical types N, C, E, A introduced in the last chapter will be discussed in the next section.

As a conclusion, we should point out that

1. The result for input w of a reference machine is a leaf labelled binary tree of height $c(w) \leq h(|w|)$. Simulating one (f,g)-h reference machine by another (f^*,g^*)-h reference machine means that for each input the two leaf labelled computation trees are the same.

2. A language accepter consists of a reference machine and a computational type T.

3. The complexities are properties of the reference machine. For a fixed reference machine, if we give it different types, the languages accepted may be different, but their complexities are the same.

4. The computational type T is a partial mapping from TREE to {0,1}. But if a machine M is of type T, then for any input w, T is always defined for the computation tree ϕ_{Mw}. In other words, ϕ_{Mw} should be always in the domain of T.

5. For logical computational models, the height of the computation tree corresponds to the sequential time, whereas for the general computational models introduced in this chapter, the height of the computation tree corresponds to the choice complexity.

§41 RESTRICTIONS

1. The sequential restriction S

What is the difference and the relation between the new general definition and the old logical definition about computational types? Intuitively, the reference input corresponds to the choice sequence mentioned in Chapter 8. But, under the old definition, the machine does not "remember" the choice sequence (of course, if space g(n) is large enough, then the machine can remember the choice sequence), while under the new definition, the reference input is written on the reference input tape. It can be read repeatedly, costing neither space nor reversal.

Therefore it is natural to give the following definition:

DEFINITION 41.1 (Restriction S for TM). An (f,g)-h reference TM M satisfies Restriction S, if its reference input tape head can move only from left to right.

In other words, the reference tape head can move one square to the right (R) or remain stationary (S).

THEOREM 41.1 The reference TM of type Tn under restriction S is the non-deterministic TM. More precisely,

(1) if M is an (f,g)-NTM of choice complexity h(n) (ref. Definition 36.5), then there is an (f,g)-h reference TM of type Tn under restriction S, accepting L(M);

(2) if M is an (f,g)-h reference TM of type Tn under restriction S, then there is an (f,g)-NTM of choice complexity h(n) accepting L(M).

Proof: (1) suppose that M is an (f,g) NTM of choice complexity h(n). We construct an (f,g)-reference TM M' of type Tn under restriction S as follows: when machine M enters a non-deterministic state, machine M' moves its reference input head one square to the right. The machine M' uses δ_0 or δ_1 as its next move function according to whether the symbol read on the reference input tape is 0 or 1.

Obviously, the reversal of M' is not more than f(n), the space of M' is not more than g(n) and the choice complexity of M' is h(n).

(2) Now suppose that M is an (f,g)-h reference TM of type Tn under restriction S. We construct an NTM M' as follows: the machine M' non-deterministically guesses the symbol (0 or 1) on every square of the reference input tape of M. Thus M' simulates M. The reversal and space used by M' are not more than f and g respectively. Obviously the choice complexity of M' is h(n).

THEOREM 41.2 The reference TM's of types Tc, Te, Ta under restriction S are CTM, ETM and ATM respectively.

DEFINITION 41.2 (Restriction S). An (f,g)-h reference machine M satisfies restriction S, if it can be simulated by an (f^*, g^*)-h reference TM under restriction S.

This definition is very formal. Anyway, we can see what is a reference vector machine under restriction S, what is a reference RAM under restriction S,..., and so on, in the exercises.

II. The parallel restriction P

Since the length of choice sequence of a non-deterministic vector machine (NVM) is not more than the parallel time, we give the following definition.

DEFINITION 41.3 (Restriction P). An (f,g)-h reference machine satisfies restriction P, if its choice complexity $h(n) \leq f^*(n)$.

THEOREM 41.3 The reference TM of type Tn under restriction PS is essentially the non-deterministic VM (NVM), where restriction PS means restriction P and restriction S.

Proof: what we should prove is that the reference TM of type Tn under restriction PS and NVM can simultaneously simulate each other. But the former is by Theorem 41.1 an NTM, with choice complexity $\leq f^*$ (f is the bounding function for reversal). Therefore we should prove that this kind of NTM can simulate NVM simultaneously, and vice versa.

Without loss of generality, we can assume that this kind of NTM moves non-deterministically only when a phase begins, since we can interrupt a phase whenever there is a non-deterministic move. The total number of phases will be at most $f + h \leq f + f^* \leq f^*$, still polynomially related to f.

Because a phase (it is deterministic now) of this kind of NTM can be simulated by the VM within time $O(\log^* g) = O(f^*)$ and space $O(g^*)$, and the non-deterministic mechanism of this kind of NTM can be simulated by the non-deterministic mechanism of the NVM, we know that this kind of NTM can be simulated simultaneously by an NVM. Conversely it is easy to prove that an NVM can be simultaneously simulated by this kind of NTM.

III. Restrictions P' and S'

For the parallel restriction P there is a natural dual restriction P'.

DEFINITION 41.4 (Restriction P'). An (f,g)-h reference machine M satisfies restriction P', if its choice complexity $h \leq g^*$. We call this the spatial restriction.

LEMMA 41.1 An (f,g)-h reference machine M satisfies restriction P', iff M can be simulated by a $(g^*, 2g^*)$-h reference machine M' under restriction P.

Proof: if M' satisfies restriction P, then $h \leq g^*$, i.e., M satisfies restriction P'.

If M satisfies P' then $h \leq g^*$. By Theorem 39.2, M can be simulated by a $(g^*, 2g^*)$-h reference machine with $h \leq g^*$, i.e., under restriction P.

Inspired by this lemma, we define a dual restriction to S:

DEFINITION 41.5 (Restriction S'). An (f,g)-h reference machine M satisfies restriction S', if it can be simulated by a $(2f^*, f^*)$-h reference machine M' under restriction S.

LEMMA 41.2 If M is an (f,g)-h reference machine in one model listed in Theorem 39.1, then in any other model listed in Theorem 39.1 there is an (f^*, g^*)-h reference machine M' to simulate M. When M satisfies restriction P,P',S,S', the simulating machine M' also satisfies restriction P,P',S,S' respectively.

Proof: this is a direct conclusion of the definitions and Theorem 39.1.

THEOREM 41.6 (1) A machine satisfying P' can be simulated by a machine satisfying S.
    (2) A machine satisfying P can be simulated by a machine satisfying S'.

Proof: by Lemma 41.2, we need only consider reference TM. Suppose that M is an (f,g)-h reference TM.

(1) If M satisfies restriction P', i.e., $h \leq g^*$, then we can copy the reference input on a work tape, and use this work tape to read the reference input later. Therefore the reference input tape head need not move to the left.

It should be noticed that if the machine reads the reference input on the work tape repeatedly, the reversal may increase, while reading information on the reference input tape repeatedly will not increase reversal. Therefore we should design the simulating reference TM M' according to different cases.

191

When $g \leq f^*$ (narrow case), M' copies the reference input on its work tape and uses the work tape instead of the reference input tape. The reversal increased is not more than the sequential time, which is $\leq f^*$.

When $f \leq g^*$ (flat case), M' copies the reference input on its work tape in one phase. Then it duplicates $f^*g^* \leq g^*$ copies of the reference input on the work tape so that the machine need not change the direction of the head of this tape when reading the reference input repeatedly. In fact, the machine uses two work tapes to simulate the reference input tape of M: one contains $f^*g^* \leq g^*$ copies of the reference input (each of length h, which can be obtained in f^* reversal and g^* space). The other one is a graduated ruler consisting of $f^*g^* \leq g^*$ segments each of length h-1. If the machine M moves its reference input heads to the left, the machine M' moves its two tape heads to the right a distance h-1, by means of the graduated ruler. The work needs $O(\log f^*g^*) = O(f^*)$ reversal and $O(g^*)$ space.

In either case, M' satisfies restriction S.

(2) If M satisfies restriction P, i.e., $h \leq f^*$, then by Theorem 39.3, M can be simulated by a $(2^{f^*}, f^*)$-h reference TM M'. Thus M' satisfies restriction P'. Therefore we can assume that M' satisfies restriction S. By Definition 41.5, M satisfies restriction S'.

By this theorem, we may say that restriction P' implies restriction S and restriction P implies restriction S' as far as polynomial simulation is allowed.

We can construct many restrictions by means of AND and OR operations from the four restrictions S,S',P,P'. At least we can get the nine restrictions shown in the Fig. 41.1, where ∅ means no restriction at all. As before, PS means restriction P AND restriction S,..., and so on.

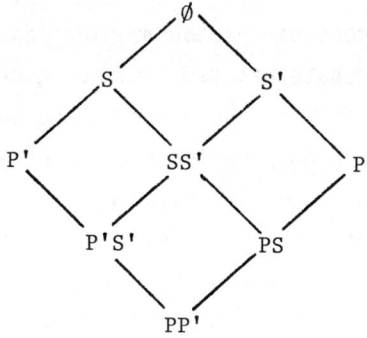

Fig. 41.1

LEMMA 41.3   P' U P = ∅.

Proof: if $f^* \geq g$ then $h \leq f^*g^* \leq f^*$, P = ∅. If $g^* \geq f$ then $h \leq f^*g^* \leq g^*$, P' = ∅. Thus P'UP = ∅.

THEOREM 41.7 The restrictions constructed by means of AND and OR from S,S', P,P' are all shown in Fig. 41.1

Proof: in fact, the nine restrictions are closed under the operation OR. For example, since PUP' = ∅, we have S'UP' = ∅, SUP = ∅, SS'UP = (SUP)(S'UP) = ∅S' = S', and so on.

In the following, by restriction R we always mean one of the nine restrictions.

THEOREM 41.8 Suppose that M is an (f,g)-h reference machine. Then the restrictions of M satisfy the following relations as far as polynomial simulation is allowed:
1. When $f \leq g^*$, ∅ = S = P', S' = SS' = P'S', P = PS = PP'.
2. When $g \leq f^*$, ∅ = S' = P, S = SS' = PS, P' = P'S' = PP'.
3. When $f \sim {}^*g$, ∅ = S = P' = S' = SS' = P'S' = P = PS = PP'.

Proof: when $f \leq g^*$, since $h \leq f^*g^*$, we have $h \leq g^*$. In other words ∅ = P'. Thus ∅ = S = P'. The others hold by the same reason.

Exercises

41.1 In the narrow case, consider vector machines and uniform aggregates. Suppose that the aggregate has two indirect addressing devices, one for input, the other for reference input, and that the indirect address for reference input can only be increased. More precisely, there are two gates A and B. Whenever the value of A is 1, the value of B is the next reference input bit. If the value of A is 0, the value of B does not change. The vector machines can be treated similarly. Prove that these models are similar to reference TM under restriction S. (In the flat case, $S = \emptyset$.)

41.2 For uniform circuits, suppose that the fan-out of each reference input gate is one and that the reference input gate connecting to a higher level has larger coding. Prove that this model is similar to reference TM under restriction S.

41.3 What is the restriction S for RAM or PRAM?

41.4 Prove Theorem 41.2 and Lemma 41.2.

§42 THE THIRD SIMILARITY THEOREM AND COMPLEXITY CLASSES

To unify all computational models and computational types we prove the following theorem.

THEOREM 42.1 (The Third Similarity Theorem). For any computational type and any restriction R, the computational models listed in Theorem 39.1 are all similar, i.e., they can simulate each other simultaneously.

Proof: by Theorem 39.1, the reference machines of any two computational models listed can simultaneously simulate each other. According to the definition, the simulation between two reference machines means that for the same input w, the computation trees are the same. Therefore the values under the mapping T (type T is a mapping) are the same. By Lemma 41.2, the simulation maintains the restriction R. Thus the models of type T under restriction R are similar.

DEFINITION 42.1 Suppose that T is a computational type, R is a restriction.

1. If (f,g) is a nice pair, then we use $TR(f^*,g^*)$ to represent the language class accepted by some (f^*,g^*)-h reference machine of type T under restriction R.

2. $TR(f^*,--) = \bigcup\limits_{g:(f,g)\text{ nice}} TR(f^*,g^*)$.

3. $TR(--,g^*) = \bigcup\limits_{f:(f,g)\text{ nice}} TR(f^*,g^*)$

4. $TR(f^*) = TR(f^*,f^*)$.

Therefore $TR(f^*,--)$ is the language class accepted in f^* reversal under type T and restriction R. $TR(--,g^*)$ is the language class accepted in g^* space under type T and restriction R. $TR(f^*)$ is the language class accepted in sequential time f^* under type T and restriction R.

For example, since we know Td is D, we can write D instead of Td, then $D(\log^* n, n^*)$ is NC; $D(n^*, \log^* n)$ is SC (in honour of S.A. Cook); $D(n^*)$ is the language class accepted in polynomial time by some deterministic machine, usually denoted by P; $Tn(n^*)$ is the language class accepted in non-deterministic polynomial time, usually denoted by NP; $D(--,n^*)$ is often denoted by PSPACE; $TnS(--,n^*)$ is often denoted as NPSPACE.

From Figure 41.1, we have the containment relations shown in Figure 42.1.

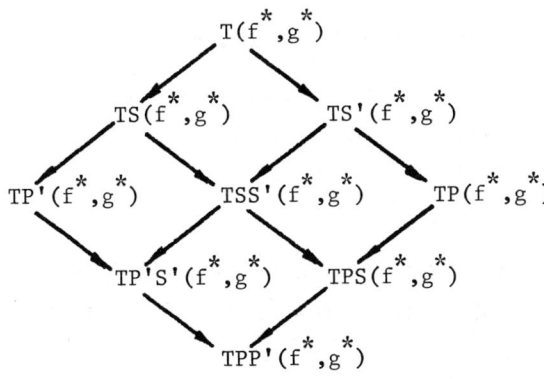

Fig. 42.1.

THEOREM 42.2 When f = g, the nine language classes shown in Fig. 42.1 are all the same.

Proof: in this case, ∅ = PP'.

This means that while in general NTM is stronger than NVM (i.e., TnPS(f^*,g^*) ⊆ TnS(f^*,g^*)), their abilities are the same if we consider only sequential time.

THEOREM 42.3 For type Td, the nine language classes shown in Figure 42.1 are all the same.

Proof: the reference TM of type Td is DTM, and DTM can be simulated by 0-reference TM of type Td. Since h(n) = 0, the machine satisfies restriction PP'. Therefore Td(f^*,g^*) = TdPP'(f^*,g^*).

THEOREM 42.4 For any computational type T, we have

(1) T(--,g^*) = TS'(--,g^*) = TP(--,g^*)

   TS(--,g^*) = TSS'(--,g^*) = TSP(--,g^*)

   TP'(--,g^*) = TP'S'(--,g^*) = TP'P(--,g^*)

(2) T(f^*,--) = TS(f^*,--) = TP'(f^*,--)

   TS'(f^*,--) = TSS'(f^*,--) = TP'S'(f^*,--)

   TP(f^*,--) = TPS(f^*,--) = TPP'(f^*,--)

Proof: this is a direct consequence of Theorem 41.8.

Therefore, if we consider only space complexity, the NTM and the NVM have the same ability (TnS(--,g^*) = TnSP(--,g^*)).

Exercises

42.1 For any restriction R and any nice pair (f,g), prove the relations shown in Figure 42.2 (the arrow means the relation ⊆).

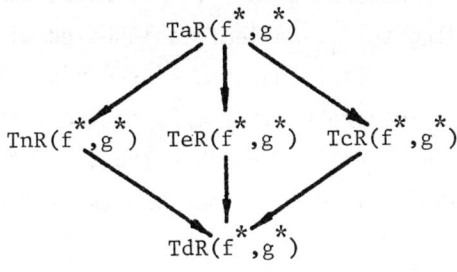

Fig. 42.2

42.2 Prove that

(1) $L \in TnR(f^*, g^*) \leftrightarrow L^C \in TcR(f^*, g^*)$

(2) $L \in TeR(f^*, g^*) \leftrightarrow L^C \in TeR(f^*, g^*)$

(3) $L \in TaR(f^*, g^*) \leftrightarrow L^C \in TaR(f^*, g^*)$.

§43 THE MAJORITY AND RANDOM TYPES

First, we define Turing machines of majority type (MTM) and random type (RTM) in the same manner as in section 36. We consider only the case when $t(n) < \infty$.

DEFINITION 43.1   A k-work tape TM of majority type (MTM) is a nine-tuple

$$M = (Q, I, \Sigma, \sqcup, \delta_0, \delta_1, q_0, F, G)$$

where $Q, I, \Sigma, \sqcup, \delta_0, \delta_1, q_0, F, G$ are used as in Definition 36.2.

For any input word w, M uses δ_0 or δ_1, each with probability 1/2, to determine its next movement, until it reaches an accepting or rejecting state. If the probability that M reaches an accepting state is $\geq 1/2$ then M accepts w, else M rejects w.

If we further assume that for any input word w, if the probability $P(w)$ that M reaches an accepting state is > 0 then the probability $P(w)$ is $\geq 1/2$, then the machine is of type random (RTM). The complexities are defined in the same way as in Definition 37.1.

Thus for every input, the computation of M is a binary tree. Each node of the tree is labelled with an ID. The left-hand son of this node is labelled with the next ID according to δ_0. The right-hand son is labelled with the next ID according to δ_1. A leaf is labelled with an accepting ID or a rejecting ID, or simply labelled with 1 or 0 respectively. We go down from the root, with the same probability 1/2, to the left-hand son or the right-hand son at any node, until we reach a leaf. If the probability that we reach a leaf with label 1 is $\geq 1/2$, then the machine accepts. Otherwise the machine rejects the input.

Notice that if the machine M is of random type, the probability that we reach a leaf with label 1 is less than 1/2 implies that there is no such leaf. There is a "gap", by which the random type is different from the majority type.

For the computation tree shown in Figure 43.1, the probability that M reaches a leaf with an accepting state is 5/8. Therefore the machine accepts the input.

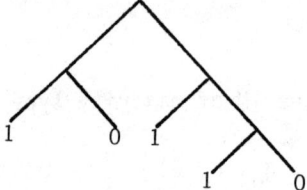

Fig. 43.1

Now we give another definition according to the point of view in this chapter.

<u>DEFINITION 43.2</u> (Type Tm and Type Tr). For a leaf labelled binary tree $\phi \in$ TREE, define

$$Tm(\phi) = \begin{cases} 1 & \text{when } \sum_x \phi(x) \cdot 2^{-|x|} \geq 1/2 \\ 0 & \text{when } \sum_x \phi(x) \cdot 2^{-|x|} < 1/2 \end{cases}$$

$$Tr(\phi) = \begin{cases} 1 & \text{when } \sum_x \phi(x) \cdot 2^{-|x|} \geq 1/2 \\ 0 & \text{when } \sum_x \phi(x) \cdot 2^{-|x|} = 0 \\ \text{undefined, otherwise.} \end{cases}$$

THEOREM 43.1   The reference machine of type Tm(Tr) under restriction S is MTM (RTM).

In the following, we consider the relations between the new computational types Tm, Tr and other computational types.

THEOREM 43.2   For any restriction R and any nice pair (f,g), we have: if $L \in TmR(f^*, g^*)$, then $L^C = (0,1)^* - L \in TmR(f^*, g^*)$.

Proof: suppose $L \in TmR(f^*, g^*)$. Then there is an (f^*, g^*)-h reference TM M of type Tm under restriction R such that L(M) = L. We construct an h+1 reference TM M' of type Tm as follows. (Obviously h+1 is log-space constructible, $h + 1 \leq f^* g^*$ and its unary and binary representations can be computed in f^* reversal and g^* space.)

We use a counter to guarantee that the first reference input bit will be read only once (when $f \leq g^*$, use a unary counter, when $g \leq f^*$, use a binary counter).

If this bit is 0, then the machine M' simulates M (taking the other reference input bits as the reference input of M). But whenever M enters an accepting state, M' enters a rejecting state; whenever M enters a rejecting state, M' enters an accepting state.

If this bit is 1, then M' reads its reference input. If the reference input are all 1's (the total length h+1 can be computed) or the second bit is 0, then it rejects, otherwise M' accepts.

Thus, the labelled binary tree $\phi_{M'w}$ is one level higher than ϕ_{Mw}. The left sub-tree of $\phi_{M'w}$ is exactly the same as that of ϕ_{Mw}, but with all labels exchanged i.e., 0 becomes 1, and 1 becomes 0. All leaves on the right left subtree have label 0. All leaves on the right right subtree have label 1, except the rightmost leaf, which has a label 0 (see Figure 43.2). Therefore $L(M') = L^c$.

Obviously, M' uses f^* reversal and g^* space. Since M is under restriction R, so is M'.

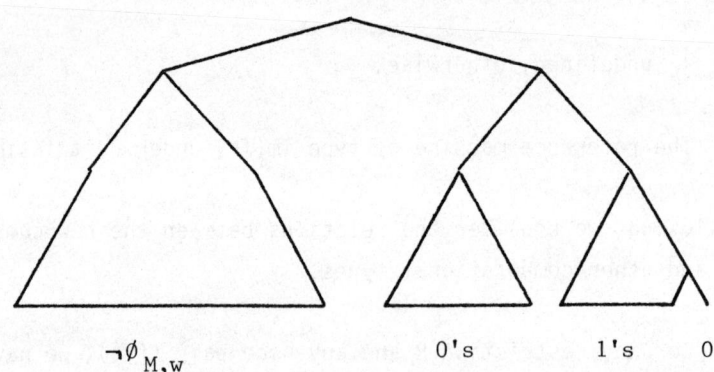

$\neg\emptyset_{M,w}$ 0's   1's   0

Fig. 43.2

THEOREM 43.3 For any restriction R and any nice pair (f,g), we have

$$TcR(f^*,g^*) \subseteq TmR(f^*,g^*).$$

Proof: for any (f^*,g^*)-h reference TM M of type Tc under restriction R, we construct an h+1 reference TM M' of type Tm as follows. Using the same technique as in the proof of the last theorem, we can guarantee that the first reference input bit will be read only once. If this bit is 0, then M' simulates M. If this bit is 1, then M' rejects. Namely:

$\phi_{M'w}(0x) = \phi_{Mw}(x)$ $(|x| = h(|w|))$

$\phi_{M'w}(1x) = 0.$

Therefore (see Figure 43.3)

$Tc(\phi_{Mw}) = 1$ iff

all leaves of ϕ_{Mw} have label 1 iff

the number of leaves with label 1 in $\phi_{M'w}$ is at least one half, iff

$Tm(\phi_{M'w}) = 1$.

Thus, $L(M) = L(M')$, the machine M' of type Tm uses f* reversal, g* space and satisfies restriction R.

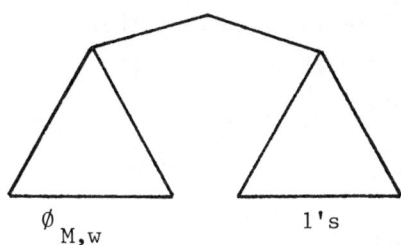

Fig. 43.3

THEOREM 43.4 For any restriction R and any nice pair (f,g), we have

$TnR(f*,g*) \subseteq TmR(f*,g*)$.

Proof: suppose $L \in TnR(f*,g*)$. Then by Exercise 42.2, $L^c \in TcR(f*,g*)$. Therefore $L^c \in TmR(f*,g*)$ by Theorem 43.3. Thus $L = L^{cc} \in TmR(f*,g*)$ by Theorem 43.2.

From Theorems 43.3, 43.4, we can see that the ability of type Tm is at least the same as that of type Tc or Tn. If we take the restriction R to be S, then the ability of MTM is at least the same as that of NTM or CTM. If we take R to be PS, then the ability of MVM is at least the same as that of NVM or CVM (though we have not defined MVM nor CVM).

THEOREM 43.5 Let (f,f) be a nice pair. Then

$Tm(f*) \subseteq Ta(f*)$.

Proof: suppose $L \in Tm(f^*,f^*)$. Then there is an (f^*,f^*)-h reference TM M of type Tm accepting L. We use ϕ_{Mw} to represent the labelled binary tree. Without loss of generality, we can assume that ϕ_{Mw} is a complete binary tree of height $h(n)$. Let $n = |w|$. Then

   $w \in L$ iff

the number of leaves with label 1 in $\phi_{Mw} \geq 2^{h(n)-1}$ iff

there exists $i \leq 2^{h(n)-1}$ such that the number of leaves with label 1 in the left subtree is $\geq i$, and the number of leaves with label 1 in the right subtree is $\geq 2^{h(n)-1} - i$.

Generally, let $P(x,i)$ be the predicate that the number of leaves with label 1 in a subtree rooted at x is $\geq i$. Obviously we have

(1) when $i > 2^{h(n)-|x|}$, $P(x,i) = 0$,

(2) when $i \leq 2^{h(n)-|x|}$,

   (a) if $|x| = h(n)$ then $P(x,i) = 1$ ($i = 0$),

$$P(x,i) = M(w,x) \ (i = 1);$$

   (b) if $|x| < h(n)$ then

$$P(x,i) = (\exists j \leq i)(P(x0,j) \wedge P(x1,i-j)).$$

By these, we can write the following recursive program to determine $P(x,i)$.

  FUNCTION $P(x,i)$

  BEGIN

    IF $i > 2^{h(n)-|x|}$ then return 0;

    IF $|x| = h(n)$ THEN

      IF $i = 0$ THEN return 1 ELSE return $M(w,x)$

    select disjunctively $j \leq i$;

    select conjunctively $a \in \{0,1\}$;

    IF $a = 0$ THEN return $P(x0,j)$ ELSE return $P(x1,i-j)$

  END

For simulating M, we need only to call $P(\Lambda, 2^{h(n)-1})$. The alternating time is $O(f*)$.

This theorem says that if we consider only sequential time, type Ta is at least as strong as type Tm.

THEOREM 43.6 For any nice pair (f,g) and any restriction R, we have $TrR(f*,g*) \subseteq TnR(f*,g*)$.

Proof: it follows directly from the definition.

Therefore Tm is between Ta and Tn (when we consider sequential time), and Tr is between Tn and Td. In practice, type Tr is very useful. If a language L is accepted by a machine M of type Tr, then in order to decide whether $w \in L$, we can run M randomly t times. Once the machine M answers "yes", we know certainly $w \notin L$. If the machine answers "no" t times, it is probably true that $w \notin L$, since in this case the probability that $w \in L$ is $\leq 1/2^t$.

Another advantage of type Tr is that the machine can be "derandomized", if we give the machine some advice. In other words, a random type machine can be simulated by a deterministic machine given proper "advice".

DEFINITION 43.3 An advice TM is a deterministic TM having one input tape, one read only advice input tape and k work tapes. The input and advice input tapes are read only. For any input length n, there is a fixed advice word a_n, which is put on the advice input tape.

In the definition, the mapping $n \to a_n$ is not necessarily recursive.

THEOREM 43.7 (Adleman). For any Tr type TM M of time t(n), there is an advice TM M' to simulate M such that the time used by M' is $O(nt(n))$ and the length of the advice word is $O(nh(n))$, where h(n) is the choice complexity.

Proof: the computation tree of M is of height h(n), therefore there are $2^{h(n)}$ leaves. If $w \in L$, then at least one half of the leaves are accepting. Otherwise all leaves are rejecting. Therefore for each w, we can find a leaf such that $w \in L(M)$ iff the leaf is accepting. Thus we can record the name of the leaf on the advice input. But there are 2^n different input words of length n. Can we find some representative leaves (not too many) in common so that w ∈ L iff at least one of the representative leaves is accepting?

The size of the set W = {w|w ∈ L(M), |w| = n} is at most 2^n. For each word w in the set, there are at least $2^{h(n)-1}$ accepting leaves. Therefore there is a representative leaf which "represents" one half of the words in W. Thus there are at most 2^{n-1} members in W left. We select another representative leaf so that there are at most 2^{n-2} members in W left, ... and so forth. In this way we can select n+1 representative leaves so that w ∈ L iff on the computation tree at least one of them is accepting.

We record the names of the n+1 representative leaves on the advice input. The length of each is h(n). Therefore the total advice length is O(nh(n)). For each representative leaf we should simulate M once. Thus the total deterministic time is O(nt(n)).

Exercises

43.1 Prove Theorem 43.1

43.2* The symmetry type is introduced by Lewis and Papadimitriou. A symmetric TM (STM) is an NTM (according to Section 35, at each step there may be more than two selections) in which the relation "⊢" is symmetric, i.e., for any two ID's D_1 and D_2, $D_1 \vdash D_2$ iff $D_2 \vdash D_1$. Suppose that (f,g) is a nice pair and h ≤ f*g* is log-space constructible. We further assume that after h moves the machine will automatically stop If along any path the reversal is ≤ f and space is ≤ g, then we call it an (f,g) symmetry TM of choice complexity h(n). Find a type Ts according to Definition 40.3 such that for any (f,g) symmetry TM of choice complexity O(h(n)) there is an (f*,g*)-O(h) TM of type Ts under restriction S simulating it and vice versa.

43.3 Prove that Tm(f*,f*) ⊆ D(-,f*) = Ta(f*) by using depth first search on the computation tree.

43.4 Prove Theorem 43.3 without using the technique in the proof of Theorem 43.2, i.e., computing the binary or unary representation of h(n). Therefore M' has the same reversal and space complexity as M. [Hint: when the reference input tape of M moves to the left, the reference input of M' moves left two squares and then moves right one square.]

§44 COMPLETE PROBLEMS

DEFINITION 44.1 Suppose that L_1 and L_2 are two languages over alphabet $\{0,1\}$, and M is a TM (a transducer) of space complexity $O(\log n)$. If, for any input w, it holds that

$$w \in L_1 \quad \text{iff} \quad M(w) \in L_2$$

then L_1 is log-space reducible to L_2.

In this definition, if M is a TM of time complexity $O(n^*)$ instead of log-space complexity, then L_1 is polynomial time reducible to L_2. Obviously if L_1 is log-space reducible to L_2 then it is polynomial time reducible to L_2.

DEFINITION 44.2 Suppose that C is a language class and L is a language. If any language in C is log-space (polynomial time) reducible to L, then L is C-hard under log-space (polynomial time) reduction. If furthermore L belongs to C, then L is C-complete under log-space (polynomial time) reduction.

THEOREM 44.1 Suppose that T_1 and T_2 are two computational types, that L is $T_1(n^*)$-hard under polynomial time reduction, and that $L \in T_2(n^*)$. Then $T_1(n^*) \subseteq T_2(n^*)$.

Proof: because L is $T_1(n^*)$-hard under polynomial time reduction, for any language $L_1 \in T_1(n^*)$ there is a polynomial time determinisitc TM M_1 satisfying

$$w \in L_1 \quad \text{iff} \quad M_1(w) \in L \quad \text{for all} \quad w \in \{0,1\}^*.$$

Since $L \in T_2(n^*)$, there is a polynomial time TM M_2 of type T_2 satisfying that $M_2(M_1(w)) = 1$ iff $M_1(w) \in L$ iff $w \in L_1$. The TM $M_1 \cdot M_2$ of type T_2 accepts L_1. The total sequential time of $M_1 \cdot M_2$ is not more than $n^* + |M_1(w)|^* = n^* + |w|^{**} = n^*$.

THEOREM 44.2 Suppose that T_1 and T_2 are two computational types, that (f,g) is a nice pair such that $f(n^*) = O((f(n))^*)$ and $g(n^*) = O((g(n))^*)$, that L is $T_1(f^*,g^*)$-hard under log-space reduction, and that $L \in T_2(f^*,g^*)$. Then

205

$$T_1(f^*,g^*) \subseteq T_2(f^*,g^*).$$

Proof: as an exercise.

In the following, we give an NP-complete problem under log-space reduction, the satisfiable problem of Boolean conjunctive normal form proposed by S.A. Cook.

SAT : Input: a coding of a Boolean expression in conjunctive normal form.
Problem: whether or not we can assign to each Boolean variable a value 0 or 1 (false or true) such that the value of the Boolean expression is true?

Suppose that a language $L \in \{0,1\}^*$ belongs to NP. Then there is a non-deterministic uniform circuit of polynomial size to accept it. The circuit has n input gates, some reference input gates, a special output gate g, and many other gates. The total number of gates is not more than n^*. The circuit accepts input $w = a_1 a_2 \ldots a_n$ ($a_i = 0,1$) iff when we assign to the n input gates the values $a_1, a_2, \ldots, a_n$ in order, there is a suitable system of assignments to the reference input gates such that the value of the special output gate g is 1. The coding of the circuit is a collection of information segments in the following form:

(type, name, input1, input2)

We use $1, 2, \ldots, n$ as the codings of input gates, $n+1, \ldots, m$ as the codings of reference input gates, $m+1, \ldots$ as the codings of other (ordinary) gates. In the information segment, "name" is the coding for an ordinary gate, hence $\geq m+1$, "input1" and "input2" are the codings of the two gates to which the two input lines of "name" are connected; "type" is one of $\wedge$ or $\neg$ (ref. Exercise 30.8). If "type" is $\neg$, then "input1" and "input2" are the same. We further assume that "name_g" is the coding for the special output gate g. The whole coding of the circuit is denoted by C_1, which is log-space constructible.

Then we rewrite C_1 by rewriting each its information segment as follows:
(1) When "type" is $\wedge$, we rewrite ($\wedge$, name, input1, input2) to (name = input1∧input2), then to

($\neg$ name $\vee$ (input1 $\wedge$ input2))($\neg$input1$\vee$ $\neg$input2$\vee$name),

$(\neg\text{name} \vee \text{input1})(\neg\text{name} \vee \text{input2})(\neg\text{input1} \vee \neg\text{input2} \vee \text{name})$.

(2) When "type" is $\neg$, we rewrite $(\neg, \text{name}, \text{input1}, \text{input1})$ to

$(\text{name} = \neg \text{input1})$,

then to

$(\neg \text{name} \vee \neg \text{input1})(\text{input1} \vee \text{name})$.

Taking the conjunction of all these information segments, concatenating further (name_g) on it, we obtain a coding C_2. If we think of each integer as a Boolean variable, then C_2 is a coding of a Boolean expression in conjunctive normal form. Substituting $a_1, a_2, \ldots, a_n$ (0 or 1, the input bits) for the variables represented by integers $1, 2, \ldots, n$, simplifying the expression according to the rules $\neg 1 = 0$, $\neg 0 = 1$, $y \vee 1 = 1$, $y \vee 0 = y$, we obtain coding C_3. Obviously $w = a_1 a_2 \ldots a_n \in L$ iff the Boolean expression represented by C_3 can be satisfied. Rewriting C_3 in a binary coding form, we obtain C_4. The coding C_4 is log-space constructible and of length $O(n^*)$ (ref. Theorem 17.4).

Therefore there is a log-space DTM M satisfying

(1) $M(w) = C_4$,

(2) $w \in L$ iff the Boolean expression represented by C_4 can be satisfied.

Thus, we have proved the following theorem.

THEOREM 44.3 (Cook) The problem SAT is NP-complete. Furthermore, the satisfiable problem for conjunctive normal form in which each term is a disjunctive form of not more than three literals (3-SAT) is NP-complete under log-space reduction.

Notice that in the proof we have not used any special property about non-determinism. We can prove the following theorems in the same way.

THEOREM 44.4 The problem of evaluating a conjunctive normal form with universal and existential quantifiers is $Ta(n^*)$ complete under log-space reduction.

The problem is to evaluate an expression of the following form

$$\forall x_1 \exists x_2 \forall x_3 \exists x_4 \ldots P(x_1, x_2, \ldots, x_k),$$

where $P(x_1, x_2, \ldots, x_k)$ is a Boolean expression in conjunctive normal form with variables $x_1, x_2, \ldots, x_k$ (even when we restrict the number of literals in each term to ≤ 3).

Since the language class accepted in polynomial alternating time is the language class accepted in polynomial space, we have

THEOREM 44.5 The quantified 3-CNF problem in Theorem 44.4 is PSPACE complete under log-space reduction.

It is not difficult to prove the following theorems:

THEOREM 44.6 The problem of determining whether the number of solutions for a Boolean expression in 3-conjunctive normal form is even, is $Te(n^*)$ complete under log-space reduction.

THEOREM 44.7 The problem of determining whether the number of different solutions for a Boolean expression in 3-conjunctive normal form is not less than a binary number k, is $Tm(n^*)$ complete under log-space reduction.

Proof: it is easy to see that the problem is $Tm(n^*)$ hard. By the same method used in Theorem 43.2 and Theorem 43.3, we can prove that the problem is in $Tm(n^*)$.

Exercises

44.1 Let L_1 be an NP-hard language, T be a log-space TM such that

$$w \in L_1 \text{ iff } T(w) \in L_2.$$

Prove that L_2 is NP-hard.

44.2 (Karp) Reduce the problem SAT to the following problem:
CLIQUE: does an undirected graph have a clique (a complete subgraph) of size k (given the coding of the graph and the binary representation of k as inputs)?

44.3 (Karp). A graph G is k-colourable if there exists an assignment of the integers 1,2,3,...,k, called "colours" to the vertices of G such

that no two adjacent vertices are assigned the same colour. Prove that the following problem is NP-complete by reducing 3-SAT to it:
COLOURABILITY: given the coding of an undirected graph G and the binary representation of a natural number k, is G k-colourable?

10 The duality between parallel time and space

Perhaps the reader has noticed that all the theorems in this book are symmetric: parallel time is dual to space, while sequential time is dual to itself. Roughly speaking, anything that is true for reversal is true for space within polynomial simulations, and vice versa. Though we cannot prove a general duality principle, we can certainly give a series of metatheorems in a symmetric form.

We are not sure whether there is a counterexample to the duality principle when the bounding functions form a nice pair. But we are sure that it is useful to consider problems from this point of view: if a theorem is true for space then we can guess a dual theorem for reversal and try to prove it, and vice versa. It is often true.

§45 THE BOUNDARY THEOREM AND THE TRANSFORM THEOREM

THEOREM 45.1 (The Boundary Theorem). Suppose that f is a nice function, R is a restriction and T is a computational type. Then we have

(1) $TR(f*,--) = TR(f*,2^{f*})$,

(2) $TR(--,f*) = TR(2^{f*},f*)$.

Proof: this is a direct conclusion of the definition of a nice pair of functions.

Notice that in the theorem, $TR(f*,2^{f*})$ and $TR(2^{f*},f*)$ are not defined. We should define

$$TR(f*,2^{f*}) = \bigcup_k TR(f^k,2^{f^k})$$

$$TR(2^{f*},f*) = \bigcup_k TR(2^{f^k},f^k).$$

It should be pointed out again that we consider only the case when the bounding functions form a nice pair. If it is not a nice pair, the similarity theorem may not be true.

THEOREM 45.2 (The Transform Theorem) Suppose that f is a nice function and T is a computational type. Then

(1) $T(f*,--) = T(--,f*)$

(2) $TP(f*,--) = TP'(--,f*)$

(3) $TS'(f*,--) = TS(--,f*)$

Proof: (1) Suppose that $L \in T(f*,--)$. Then there is an $(f*,g*)$-h reference machine M of type T accepting L. By Theorem 39.3 there is a $(2^{f*},f*)$-h reference machine M' simulating M. That is, for the same input the labelled binary trees for M and M' are the same. Therefore the machine M' of type T accepts L. Thus $L \in T(--,f*)$.

Conversely, if $L \in T(--,f*)$, by Theorem 39.2, we can prove $L \in T(f*,--)$ in the same way.

(2) Suppose that $L \in TP(f*,--)$. Then there is an $(f*,g*)$-h reference machine M of type T under restriction P, accepting L. By Theorem 39.3, there is a $(2^{f*},f*)$-h reference machine M' simulating M. Therefore the machine M' of type T accepts L. Since $h \leq f*$, the machine M' satisfies restriction P'. Thus $L \in TP'(--,f*)$.

Conversely, in the same way we can prove that $TP'(--,f*) \subseteq TP(f^*,--)$.

(3) Suppose that $L \in TS'(f*,--)$. Then there is an $(f*,g*)$-h reference machine M of type T under restriction S', accepting L. By the definition of restriction S', there is a $(2^{f*},f*)$-h reference machine M' under restriction S simulating M. Therefore the machine M' of type T accepts L. Thus $L \in TS(--,f*)$.

Conversely, if $L \in TS(--,f*)$ then there is a $(g*,f*)$-h reference machine M of type T under restriction S, accepting L. By Theorem 39.2, there is an $(f*,2^{f*})$-h reference machine M' simulating M. Therefore the machine M' of type T accepts L. Since M simulates M', by the definition of restriction S', the machine M' satisfies restriction S'. Thus $L \in TS'(f*,--)$.

Especially taking type T to be Td, we have

$Td(f*,--) = Td(--,f*)$.

In other words, deterministic space corresponds to deterministic parallel time. If there is a highly parallelized algorithm then there is a highly space-saving algorithm, and vice versa.

People are familiar with sequential computation. There are many mature sequential algorithms. The last theorem tells us that whenever we have a highly space-saving sequential algorithm, we can get a highly parallelized algorithm.

§46 THE SYMMETRY THEOREM AND RESTRICTION THEOREM

<u>THEOREM 46.1</u> Suppose that (f,g) is a nice pair. Then

$$Tn(f*,g*) = Tn(g*,f*) = Tn((fg)*).$$

Proof: since the formula is symmetric, we need only to prove that

$$Tn(f*,g*) = Tn((fg)*).$$

Obviously we have

$$Tn(f*,g*) \subseteq Tn((fg)*,(fg)*) = Tn((fg)*).$$

Therefore we need only to prove that

$$Tn((fg)*) \subseteq Tn(f*,g*).$$

Suppose that $L \in Tn((fg)*) = TnS((fg)*)$. Then there is a non-deterministic TM accepting L in time $O((fg)*)$. We construct an $(f*,g*)$-$(fg)*$ reference TM M' of type Tn to accept L.

Firstly, we define a "correct" computation of M for input w to be a string

$$D_0 \# D_1 \# D_2 \# \ldots \# D_t \quad t = (fg)*, \quad |D_0 \# D_1 \# D_2 \# \ldots \# D_t| \leq (fg)* \qquad (46.1)$$

where each D_i is an ID of M satisfying:

(1) D_0 is the initial ID of M for input w.

(2) $D_i \vdash D_{i+1}$ $(i = 0,1,\ldots,t-1)$. If D_i is an accepting ID then $D_{i+1} = D_i$.

If the last ID D_t in the correct computation is an accepting ID, then we call (46.1) an accepting computation. If D_t is a rejecting ID, then we call (46.1) a rejecting computation.

Suppose that the whole input of M' is (w,x), i.e., the input is w, the reference input is x. We prove this theorem by cases.

In the narrow case (g ≤ f*, (fg)* ≤ f*), the machine M' uses $O(\log f*g*)$ = $O(g*)$ space to check if x is an accepting computation of M for input w. If it is, then M' enters an accepting state. Otherwise it rejects. Obviously the machine M' uses time $O(f*)$. Therefore, it uses $O(f*)$ reversal and $O(g*)$ space.

In the flat case (f ≤ g*, (fg)* ≤ g*), the machine can check if x is an accepting computation of M of length ≤ g* in f* reversal and $O(g*)$ space.

THEOREM 46.2 Suppose that (f,g) is a nice pair. Then

$$Ta(f*,g*) = Ta(g*,f*) = Ta((fg)*).$$

Proof: we need only to prove that

$$Ta((fg)*) \subseteq Ta(f*,g*).$$

Suppose that L ∈ Ta((fg)*). Since there is no difference between the unrestricted case and the restricted case when we consider only sequential time (ref. Theorem 42.2), we can assume that there is a standard ATM M (see Corollary 37.1) of time complexity $O((fg)*)$, accepting L. The machine M starts from a conjunctive state, goes into a disjunctive state, goes back to a conjunctive state,... and so on alternately.

Now we construct a TM M' of type Ta to simulate M. The first reference input bit is a conjunctive bit, the second is a disjunctive bit,... and so on alternately.

The machine accepts the input iff there is a sequence D_0 which is the initial ID, such that for all $a_0 \in \{0,1\}$ there is a sequence $D_1 \# D_2$ such that $D_0 \vdash D_1$ by the next move function δ_{a_0} and $D_1 \vdash D_2$ such that for all $a_1 \in \{0,1\}$ there is a sequence $D_3 \# D_4 \ldots$. Therefore we can define a correct computation as a sequence of the following form:

$$D_0 \# a_0 D_1 \# D_2 \# a_1 D_3 \# D_4 \# a_2 D_5 \# \ldots \# D_{2k} \# a_k D_{2k+1} \# \ldots$$

where each D_i is an ID of M satisfying

(1) D_0 is the initial ID,

(2) $D_{2k} \vdash D_{2k+1}$ according to the next move function δ_{a_k} ($a_k = 0,1$), D_{2k} is conjunctive, and $D_{2k-1} \vdash D_{2k}, D_{2k-1}$ is disjunctive.

We use the reference input x to simulate the correct computation. Notice that all a's are conjunctive, the others are all disjunctive. Therefore we should use the even bits of x as the symbols in D_i and #, ignoring all odd bits in between, and use odd bits of x for a's. In the same way as in the proof of the last theorem, in the narrow case, the machine M' uses $O(\log f*g*)$ = $O(g*)$ space and $f*$ reversal to check whether x is an accepting computation for input w. If it is, then it accepts. Otherwise it rejects. In the flat case, the machine M' can check whether x is an accepting computation of M for input w, in $O(f*)$ reversal and $O(g*)$ space.

THEOREM 46.3 Suppose that (f,g) is a nice pair. Then

$$Tc(f*,g*) = Tc(g*,f*) = Tc((fg)*).$$

Proof: this can be obtained from the mutual complementary property of Tn and Tc. Suppose that $L \in Tc((fg)*)$. Then $L^c = \{0+1\}*-L \in Tn((fg)*) = Tn(f*,g*)$. Therefore

$$L = L^{cc} \in Tc(f*,g*).$$

The other relations can be obtained by the same way.

THEOREM 46.4 Suppose that (f,g) is a nice pair. Then

$$Tm(f*,g*) = Tm(g*,f*) = Tm((fg)*).$$

Proof: the proof is similar to that of Theorem 46.1. The simulating machine M' checks whether its reference input x is a correct computation of M, the machine being simulated. If it is not, then accepts with probability 1/2. If it is, then M' accepts iff M accepts. Obviously, the probability that the reference machine M' accepts $\geq 1/2$ iff the probability that the reference machine M accepts $\geq 1/2$.

THEOREM 46.5 Suppose that (f,g) is a nice pair. Then

$$Te(f*,g*) = Te(g*,f*) = Te((fg)*).$$

Proof: the proof is left as an exercise.

COROLLARY 46.1 Suppose that (f,g) is a nice pair. Then
1. $Ta(f*,--) = Ta(--,f*) = Ta(2^{f*})$.
2. $Tn(f*,--) = Tn(--,f*) = Tn(2^{f*})$.
3. $Tc(f*,--) = Tc(--,f*) = Tc(2^{f*})$.
4. $Te(f*,--) = Te(--,f*) = Te(2^{f*})$.
5. $Tm(f*,--) = Tm(--,f*) = Tm(2^{f*})$.

For type Td (deterministic type), we do not know if there is a similar equality. We only have

$$Td(f*,--) = Td(--,f*) \subseteq Td(2^{f*}).$$

We do not know whether $Td(f*,g*) = Td(g*,f*)$. Especially, when $f(n) = \log n$, $g(n) = n$, $Td(\log*n, n*)$ is NC (see Definition 18.3). $Td(n*, \log*n)$ is SC. Whether or not SC = NC is still an open problem.

We cannot prove similar results for type Tr either.

THEOREM 46.6 (Restriction Theorem for Alternating) Suppose that f is a nice function. Then we have

$$TaP(f*,--) = TaP'(--,f*) = Td(f*,--) = Td(--,f*).$$

Proof: obviously, we have

$$Td(f*,--) = Td(--,f*) = TdP'(--,f*) \subseteq TaP'(--,f*) = TaP(f*,--).$$

Conversely, suppose that $L \in TaP'(--,f*)$. Then there is a $(2^{f*},f*)$-h reference TM M of type Ta, accpeting L with $h \leq f*$.

For the labelled binary tree of M for input w, we compute the value at node x by the following recursive program:

```
FUNCTION  VALUE(w,x)
BEGIN
   IF |x| = h(|w|) THEN return M(w,x);
   IF |x|<h(|w|)   THEN
      BEGIN
         a := VALUE(w,x0)
         b := VALUE(w,x1)
         IF |x| = 0 (mod 2) THEN return a∧b ELSE return a∨b
      END
END
```

In order to compute the value of the root, we call VALUE (Λ,w). The recursive depth is h ≤ f*. O(h) space is used to store x. To compute M(w,x), f* space is enough. The total space used is O(f*). Therefore L ∈ Td(--,f*).

COROLLARY 46.2 (Restriction Theorem) For any nice function f, we have

$$TnP(f^*,--) = TcP(f^*,--) = TeP(f^*,--) = TmP(f^*,--) = TrP(f^*,--) = Td(f^*,--)$$

$$= TnP'(--,f^*) = TcP'(--,f^*) = TeP'(--,f^*) = TmP'(--,f^*) = TrP'(--,f^*)$$

$$= Td(--,f^*).$$

The restriction P is the parallel restriction. The restriction P' is the spatial restriction. This corollary states that the parallel restriction restricts the (non-deterministic, alternating,...) parallel time (to be the same as the deterministic time), and the spatial restriction restricts the (non-deterministic, alternating,...) space (to be the same as the deterministic space).

The restrictions S and S' are weaker than P and P' respectively. For S and S' we have the following theorems:

THEOREM 46.7 Suppose that f is a nice function. Then we have

$$TnS'(f*,--) = TnS(--,f*) = Td(f*,--) = Td(--,f*)$$

$$TcS'(f*,--) = TcS(--,f*) = Td(f*,--) = Td(--,f*).$$

Proof: by Savitch's theorem (Theorem 35.3) and Theorem 45.2,

$$TnS'(f*,--) = TnS(--,f*) \subseteq Td(--,f*).$$

The second line comes from the first line and a similar argument as in Theorem 46.3.

THEOREM 46.8 (Chandra and Stockmeyer) Suppose that f is a nice function. Then we have

1. $Ta(f*) = Td(f*,--) = Td(--,f*)$.
2. $TaS'(f*,--) = TaS(--,f*) = Td(2^{f*})$.

Proof: directly from Theorems 38.1 - 38.4.

Exercise

46.1 Prove Theorem 46.5.

§47 THE COMPLEXITY OF FORMAL PROOF

The axiomatization method and Gödel's incompleteness theorem have a profound influence on mathematics and logic. During the last half century, mathematicians have built up a theoretical proof theory. But it is not enough to study only theoretical provability. We need also to study the realistic provability - the complexity of a proof - in order to clarify some basic properties of proofs and to speed up the process of proving theorems.

From a computational point of view, the main way to speed up is by parallelization. Can any computational problem be highly parallelized? It is unlikely. But for a formal mathematical proof, in this section, we get a definite answer: if a theorem has a proof of length L (sequential time), then it has a proof of depth O(log L) (parallel time).

Any formal proof system is a rewriting system in which we can start from certain axioms or axiom schemes to obtain new formulas by certain rules of inference. We want to give these a general form. It is important (a) that the formal system should be general enough and (b) that any inference step should remain elementary. Since it is a rewriting system, we can use a non-deterministic Turing machine to perform any inference step within a constant reversals.

DEFINITION 47.1 A normal system is a non-deterministic Turing machine. It has k two-way read only input tapes, k two-way read only auxiliary input tapes, k work tapes, a one-way write only output tape and a one-way write only auxiliary output tape. Its input, output and work alphabet sets are all Σ. Its total number of reversals of all tape heads (including input and auxiliary input heads) is bounded by a fixed constant r. It will always stop by accepting or rejecting. If the content of the i-th input tape is w_i and the content of the i-th auxiliary input tape is w_i', then when the machine accepts, the content in the output tape (auxiliary output tape) is called the direct conclusion (direct auxiliary conclusion) from premises $w_1, w_2, \ldots, w_k$ and auxiliary premises $w_1', w_2', \ldots, w_k'$.

DEFINITION 47.2 Let N be a normal system on Σ. Then (1) the null string Λ of length 0 is a theorem and is an auxiliary formula, (2) if w (or w') is a direct conclusion (or auxiliary conclusion) from theorems $w_1, \ldots, w_k$ (as premises) and auxiliary formulas $w_1', \ldots, w_k'$ (as auxiliary premises), then w (or w') is a theorem (or auxiliary formula), (3) all theorems and auxiliary formulas can be obtained by using (1) and (2).

DEFINITION 47.3 Two normal systems are equivalent if their theorem sets are the same.

EXAMPLE 47.1 Propositional calculus. In this system, there are two propositional variables a and b, two propositional connectives $\neg$ and $\rightarrow$, and two parentheses (,). There are three axiom schemes:

(1) $(A \rightarrow (B \rightarrow A))$,
(2) $((A \rightarrow (B \rightarrow C)) \rightarrow ((A \rightarrow B) \rightarrow (A \rightarrow C))$,
(3) $(((\neg A) \rightarrow (\neg B)) \rightarrow (B \rightarrow A))$,

and one rule of inference (*modus ponens*)

$$\frac{(A \to B), A}{B}$$

where A,B,C stand for some syntactically correct formulas.

To explain that it is an example of our normal system, we construct a normal system N (a non-deterministic Turing machine) such that its theorem set is exactly that of tautologies, its auxiliary formula set is exactly that of all syntactically correct formulas. N performs one of the following tasks non-deterministically:

(1) Writes a string (a) or (b) on the auxiliary output tape.

(2) If one auxiliary input is string A, then writes ($\neg$A) or (A) on the auxiliary output tape.

(3) If the auxiliary input strings are A and B, then writes (A $\to$ B) on the auxiliary output tape.

(4) If the auxiliary input strings are A and B, then writes (A $\to$ (B $\to$ A)) or ((($\neg$A) $\to$ ($\neg$B)) $\to$ (B $\to$ A)) on the output tape.

(5) If the auxiliary input strings are A,B and C, then writes ((A $\to$ (B C)) $\to$ ((A $\to$ B) $\to$ (A $\to$ C))) on the output tape.

(6) If the input strings are A and (A $\to$ B), then writes B on the output tape.

Items (1), (2), (3) above construct syntactically correct formulas; Items (4) and (5) correspond to the three axiom schemes; Item (6) corresponds to *modus ponens*. In order to perform each item, the total number of reversals of all tape heads of N is not more than nine.

It is not difficult to verify that Post system, various propositional calculi and predicate calculi are all normal systems. To argue that every inference step in a normal system is still elementary, we prove the following lemma.

<u>LEMMA 47.1</u> Let w be a direct conclusion or auxiliary one from premises $w_1, \ldots, w_k$ and auxiliary premises $w_1', \ldots, w_k'$. Then

$$|w| \leq C \left(|w_1| + \ldots + |w_k| + |w_1'| + \ldots + |w_k'| \right),$$

where $|w|$ is the length of w, C is a constant dependent only on the normal system.

Proof: in the normal system, the computation of the non-deterministic Turing machine N can be divided into not more than r stages. Within each stage, no tape head changes its move direction. Since there are only a finite number of inner states and a finite number of characters in Σ, within a constant number of steps one of the tape heads must go one square ahead. Otherwise the machine would enter an endless cycle. In other words, within C_1 steps the machine must pass a square which has been used. Let $S(i)$ be the total length of the input tapes and work tapes after the i-th stage, and $t(i)$ be the total number of steps in the i-th stage. Then we have

$$t(i) \leq C_1 S(i-1),$$

$$S(i) \leq t(i) + S(i-1) \leq (C_1 + 1)S(i-1).$$

Hence, it follows that

$$|w| \leq \sum_{1}^{r} t(i) \leq C(|w_1| + \ldots + |w_k| + |w_1'| + \ldots + |w_k'|),$$

where C is a constant independent of w. This is not only true for w, but also true for any intermediate result on the work tapes.

Therefore, each direct inference of the normal system can be performed in $\leq r$ times of scanning and rewriting by a non-deterministic Turing machine. The length of any intermediate result and the conclusion is not more than a constant $\times$ the total length of the inputs.

DEFINITION 47.4 Let N be a normal system. A proof in N is a sequence of formulas separated by ","s, in which each formula is a theorem or an auxiliary formula such that each one is either a null formula or a direct conclusion (in this case it is a theorem) or a direct auxiliary conclusion (in this case it is an auxiliary formula) of some of its predecessors. The last one must be a theorem. It is the conclusion of the proof. The sequence of formulas is called a proof of this theorem. The length of the proof is the total number of characters in the sequence. The proof length of a theorem is the length of its shortest proof.

DEFINITION 47.5 In a proof, the depth of a null string is 1. If a theorem or a formula is a direct or auxiliary direct conclusion of a set of premises, then its depth is 1 plus the maximum of the depths of these premises. (If there are several ways to define the depth of a formula in a proof, we take the smallest one.) The depth of a theorem is the minimum depth of all its proofs.

Another kind of measure of proof complexity is the width of a proof. Given an arbitrary string, we can ask if it is a proof in N. To check, put the string on the read only input tape of a deterministic Turing machine M. The width is the sum of the work space of M and the logarithm of the input length. We call M a check system for N.

DEFINITION 47.6 Let N be a normal system, M be a check system for N, and $w \in (\Sigma \cup \{,\})^*$ be a string. The sum of $\log |w|$ and the work space of M on w is called the width of w (relative to M). The width of a theorem is the minimum width of all its proofs.

Thus, width is always relative to a check system M, although this is not always stated for the sake of convenience.

Let T be a k-tape non-deterministic Turing machine, Q be the set of its inner states, and Σ be its work alphabet set including space character ␣. Then a partial instantaneous description (PID) is a word of the following form:

$$x_1 q y_1 \Delta x_2 q y_2 \Delta \ldots \Delta x_k q y_k,$$

where $\Delta \notin \Sigma \cup Q$ is a separator, $q \in Q$, $x_i y_i \in \Sigma^*$.

It represents that at one moment, T is in a state q, $x_i y_i$ is a subword of the content on the i-th tape, the i-th tape head scans the leftmost symbol of y_i. Notice that $x_i y_i$ may not be the whole content of the i-th tape, but only a part of the content close to the head.

In order to describe a simple action of T, we can write down a set of basic productions (BPS) in the following form:

$$\alpha_1 \Delta \alpha_2 \Delta \ldots \Delta \alpha_k \to \beta_1 \Delta \beta_2 \Delta \ldots \Delta \beta_k, \qquad (47.1)$$

where each pair $\alpha_i \to \beta_i$ has one of the following three forms:

(1) $b_i q a_i \to b_i q' a_i'$,

(2) $b_i q a_i \to b_i a_i' q'$,

(3) $b_i q a_i \to q' b_i a_i'$

where $q, q' \in Q$, $a_i, b_i, a_i' \in \Sigma$. The meaning of (47.1) is that if machine T is in state q, its i-th tape head scans symbol a_i, $i = 1, 2, \ldots, k$, then T may take the following actions: change the symbol scanned from a_i to a_i'; change its inner state from q to q'; leave its i-th tape head in the same position (in case 1); move its i-th tape head one square right (in case 2); or move the i-th tape head one square left (in case 3).

The set BPS of all basic productions is finite. The action of machine T can be characterized by the set PPS of partial productions, which we define as follows.

(1) BPS $\subseteq$ PPS.

(2) If both $I \to J$ and $J \to K$ belong to PPS, then so does $I \to K$.

(3) If $x_1 q y_1 \Delta \ldots \Delta x_k q y_k \to y_1' q' y_1' \Delta \ldots \Delta x_k' q' y_k'$ belongs to PPS, and $u_1, \ldots, u_k, v_1, \ldots, v_k \in \Sigma^*$, then $u_1 x_1 q y_1 v_1 \Delta \ldots \Delta u_k x_k q y_k v_k \to u_1 x_1' q' y_1' v_1 \Delta \ldots \Delta u_k x_k' q' y_k' v_k$ belongs to PPS.

(4) Any element in PPS can be obtained by using (1), (2) and (3).

Informally, (1) describes one step of T's action; (2) connects all actions sequentially; (3) is used to enlarge the scopes of the PID's in both sides of the partial production. It does not correspond to any action of the machine. It is not difficult to prove that if $\alpha_1 \Delta \ldots \Delta \alpha_k \to \beta_1 \Delta \ldots \Delta \beta_k$ belongs to PPS, then $|\alpha_i| = |\beta_i|$, $i = 1, 2, \ldots, k$, and $\alpha_1 \Delta \ldots \Delta \alpha_k$, $\beta_1 \Delta \ldots \Delta \beta_k$ are PID's.

Machine T starts from an initial state $q_o \in Q$. At the beginning, the input word $w \in (\Sigma - \{\sqcup\})^*$ is put on its first tape. The other tapes are all empty (filled by symbol $\sqcup$). If T enters some accepting state $q_t \in Q_t$ eventually, we say that T accepts input w. Formally speaking, if $u_1 q_o w v_1 \Delta u_2 q_o v_2 \Delta \ldots \Delta u_k q_o v_k \to x_1 q_t y_1 \Delta \ldots \Delta x_k q_t y_k$ belongs to PPS, and $u_1 \ldots u_k v_1 \ldots v_k \in \{\sqcup\}^*$, $q_t \in Q$, $x_1 \ldots x_k y_1 \ldots y_k \in \Sigma^*$, then we say that T accepts w.

Now we can construct a normal system N_2 from T such that, for any $w \in (\Sigma - \{\sqcup\})^*$, T accepts w iff w is a theorem in N_2. We will not distinguish the normal system N_2 and the corresponding non-deterministic Turing machine.

The machine N_2 performs one of the following tasks non-deterministically:

(1) It writes a basic production in BPS on the auxiliary output tape.

(2) If there are two auxiliary inputs $I \to J$ and $J \to K$, then it writes $I \to K$ on the auxiliary output tape.

(3) If there is an auxiliary input $\alpha_1 \Delta \ldots \Delta \alpha_k \to \beta_1 \Delta \ldots \Delta \beta_k$, then it selects $u_i, v_i \in \Sigma^*$ non-deterministically, and writes $u_1 \alpha_1 v_1 \Delta \ldots \Delta u_k \alpha_k v_k \to u_1 \beta_1 v_1 \Delta \ldots \Delta u_k \beta_k v_k$ on its auxiliary output tape. The words u_i, v_i should be selected such that the output length is not more than a constant (for example, 3) × the input length.

(4) If there is an auxiliary input $u_1 q_0 w v_1 \Delta u_2 q_0 v_2 \Delta \ldots \Delta u_k q_0 v_k \to J$, $u_1 \ldots u_k v_1 \ldots v_1 \in \{\sqcup\}^*$, and the inner state in J belongs to Q_t, then it writes w on its output tape (as a theorem).

<u>LEMMA 47.2</u> The set of theorems in N_2 is exactly the language accepted by T.

Proof: it suffices to prove that the auxiliary formula set of N_2 is exactly PPS. In fact, we can obtain any basic production in PPS by using (1) and can obtain any other partial production in PPS by using (2) and (3). Conversely, any auxiliary formula in N_2 belongs to PPS.

<u>LEMMA 47.3</u> Any proof in N_2 can be verified in log-space. In other words, there is a deterministic log-space Turing machine M to determine whether an arbitrary string is a proof in N_2 or not.

Proof: let $f_0, f_1, \ldots, f_n$ be the input, whose length is L_1. The machine M can check whether f_i ($i = 0, 1, \ldots, n$) is an auxiliary formula or a theorem (if there is a "$\to$" in the formula, it must be an auxiliary formula, otherwise it should be a theorem). Since there are at most two premises used in each item of N_2, M can check all $(i-1)^2$ possible pairs of premises one by one. This can be done by using two counters of length log n. When the premises are selected, it is enough for a space of $O(\log L_1)$ to check whether f_i can be obtained from them.

If $f \in \Sigma^*$ is accepted by T within L steps, then the only squares that might be used are those that are only L apart from the original position of the head. Therefore we can assume that a complete ID is of the form

$x_1qy_1\Delta\ldots\Delta x_kqy_k$ such that $|x_iqy_i| = 2L + 1$. The whole length of the complete ID is then $2k(L+1)-1$. We express the changes of the complete ID's in the computation step by step:

$$I_0 \to I_1 \to I_2 \to \ldots \to I_L. \tag{47.2}$$

Then, $I_r \to I_{r+s}$ is a partial production in PPS. Let $x_i q y_i$ be the descriptive word of I_r on its i-th tape. Since there are only s steps passed, only the symbols on the rightmost s positions of x_i, the leftmost s positions of y_i, and the position of q may be changed in the corresponding descriptive word of I_{r+s} (altogether $2s+1$ positions). We omit the other positions on both sides of $I_r \to I_{r+s}$, and obtain a reduced form $I'_r \to I'_{r+s}$, which is still in PPS and will be denoted by $P(I_r \to I_{r+s})$. We have

$$|P(I_r \to I_{r+s})| = 4k(s+1)-1 < 8ks. \tag{47.3}$$

LEMMA 47.4 For $r, s \geq 0$, the partial production $P(R_r \to I_{r+2s})$ can be obtained from $P(I_r \to I_{r+s})$ and $P(I_{r+s} \to I_{r+2s})$ by using Item (3) twice and Item (2) once in N_2.

Proof: consider that $I_r \to I_{r+s} \to I_{r+2s}$. Let $x_i q y_i$, $x_i' q' y_i'$, $x_i'' q'' y_i''$ be their descriptive words on the i-th tape, respectively. Let INTV denote the interval of length $4s+1$ centred at position q, $INTV_1$ denote the interval of length $2s+1$ centred at q, and $INTV_2$ denote the interval of length $2s+1$ centred at q'. Then, $INTV_1$ and $INTV_2$ are all inside INTV. We restrict both sides of $I_r \to I_{r+s}$ on $INTV_1$ and denote it by $(I_r \to I_{r+s}) | INTV_1$ (in fact, it is $P(I_r \to I_{r+s})$). We restrict both sides of $I_r \to I_{r+s}$ on INTV, and denote it by $(I_r \to I_{r+s}) | INTV$. Obviously, it can be obtained from $(I_r \to I_{r+s})|INTV_1$ by using Item (3). In the same way, we can obtain $(I_{r+s} \to I_{r+2s})|INTV$ from $(I_{r+s} \to I_{r+2s})|INTV_2$ (in fact, it is $P(I_{r+s} \to I_{r+2s})$) by using (3). Finally, by using Item (2), we obtain $(I_r \to I_{r+2s})|INTV$ (i.e., $P(I_r \to I_{r+2s})$) from $(I_r \to I_{r+s})|INTV$ and $(I_{r+s} \to I_{r+2s})|INTV$.

LEMMA 47.5 If input f is accepted by T within L steps, then f has a proof in N_2, whose length $\leq cL \log L$, and whose depth $\leq c \log L$, c being a constant independent of f.

Proof: Without loss of generality, suppose that $L = 2^e$. Consider the sequence of complete ID's:

$$I_0 \to I_1 \to I_2 \to \ldots \to I_L,$$

where the inner state in I_L belongs to Q_t. Consider the partial productions

$$P(I_0 \to I_1), P(I_1 \to I_2), \ldots, P(I_{L-1} \to I_L).$$

They are all basic productions, whose length $< 8k$, and whose depth is 2. Their total length is $\leq 8k \cdot 2^e$. We can obtain the following partial productions from them by Lemma 47.4:

$$P(I_0 \to I_2), P(I_2 \to I_4), \ldots, P(I_{L-2} \to I_L),$$

each of depth 4 and length $< 2 \cdot 8k$. Their total length is $8k \cdot 2^e$. From them, we can obtain

$$P(I_0 \to I_4), P(I_4 \to I_8), \ldots, P(I_{L-4} \to I_L),$$

each of depth 6 and length $< 4 \cdot 8k$. Their total length is still $\leq 8k \cdot 2^e$, and so on. Finally, we obtain $P(I_0 \to I_L)$ of depth $2(e+1)$. By using Item (4), we obtain f as a theorem. We write down all these formulas, obtaining a proof of f in N_2. The total length of the proof is $O(L \log L)$ and depth is $O(\log L)$.

LEMMA 47.6 If a theorem f has a proof of depth D in a normal system, then it has a proof of length $\leq c^D$ in the same normal system. Similarly, if a theorem f has a proof of width W (relative to a check system M), then it has a proof of length $\leq c^W$ in the same normal system, where c is a constant independent of f.

Proof: directly from Lemma 47.1 and Definition 47.6, the proof is immediate.

LEMMA 47.7 For any normal system N_1, there is a non-deterministic Turing machine T such that

(1) the language accepted by T is exactly the set of theorems in N_1;
(2) if f has a proof of length L in N_1, then T accepts f by some choice

225

sequence within time $O(L^2)$.

Proof: machine T non-deterministically guesses the length L and a proof of length L for f. The time used is $O(L)$. Suppose that the proof guessed is $f_0, f_1, \ldots, f_n$ ($f_n = f$), with notation designating which one is a theorem and which one is an auxiliary formula. Machine T can check whether f_i ($i = 0, 1, \ldots, n$) is a direct conclusion (or direct auxiliary conclusion if f_i is designated to be an auxiliary formula) of former premises. To do so, machine T non-deterministically selects not more than k premises and k auxiliary premises and records them on 2k work tapes. The time needed is $O(L)$. Then, machine T simulates one step of inference in N_1, to check whether or not f_i is a direct conclusion or direct auxiliary conclusion of these premises. By Lemma 47.1, the time to perform this step of inference is not more than a constant × the total length of all the premises, and therefore is $O(L)$. Thus, the total time to accept f is $O(nL) = O(L^2)$ for some choice of sequence.

THEOREM 47.1 For any normal system N_1, there is an equivalent normal system N_2 and a related check system M such that, for any theorem f, if the proof length of f in N_1 is L, then f in N_2 has a proof of length $O(L^2 \log L)$, depth $O(\log L)$ and width $O(\log L)$ (relative to M).

Proof: by Lemma 47.7, for any normal system N_1 there is a non-deterministic Turing machine T such that the language accepted by T is exactly the set of theorems in N_1. For T, we can construct a normal system N_2 such that the language accepted by T is exactly the set of theorems in N_2. Therefore N_2 is equivalent to N_1. Let f be a theorem in N_1 of proof length L. Then T accepts f in time $O(L^2)$. By Lemma 47.5, f has in N_2 a proof of length $O(L^2 \log(L^2)) = O(L^2 \log L)$, and depth $O(\log(L^2)) = O(\log L)$. By Lemma 47.3, this proof can be verified in logarithmic space, i.e., there is a check system M that can accept the proof within work space $O(\log(L^2)) = O(\log L)$.

From Theorem 47.1 and Lemma 47.6, we have the following theorem.

THEOREM 47.2 The logarithm of the length, the depth and the width are linearly related to each other (Figure 47.1).

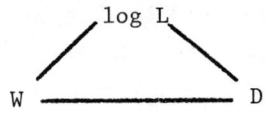

Fig. 47.1

This theorem should be stated more precisely as three theorems, one of which is the following corollary.

THEOREM 47.3 For any normal system N_1, there is an equivalent normal system N_2 and a check system M such that if a theorem f in N_1 has depth D, then its width in N_2 is $\leq$ D. Conversely, for any normal system N_2, there is an equivalent normal system N_1 and a check system M such that if a theorem f has width $\leq$ W in N_2 (relative to M), then its depth in N_1 is $\leq$ W.

Proof: the theorem follows from Theorem 47.1 and Lemma 47.6, or directly from Theorem 47.2. The constant factors can be removed by linear speed-up theorems for depth and width.

Therefore we can say: a theorem has a narrow proof iff it has a shallow proof.

References

Aho, A., J. Hopcroft and J. Ullman (1974) The Design and Analysis of Computer Algorithms. Addison-Wesley, Reading, Mass.

Aho, A., J. Hopcroft and J. Ullman (1983) Data Structures and Algorithms. Addison-Wesley, Reading, Mass.

Barzdin, J.M. and J.J. Kalnin (1974) A universal automaton with variable structure. Automatic Control and Computing Sciences 6.

Blum, M. (1967) A machine independent theory of the complexity of recursive functions. J.ACM 14:2, 322-336.

Book, R.V. (1972) On languages accepted in polynomial time. SIAM J. Comput. 1;4, 281-287.

Borodin, A. (1977) On relating time and space to size and depth. SIAM J. Comput. 6:4.

Brent, R.P. and H.T. Kung (1980) The chip complexity of binary arithmetic. 12th ACM Symposium on Theory of Computing, pp. 190-200.

Chandra, A. and L. Stockmeyer (1976) Alternation. In 17th Ann. IEEE Sympos. Found. Comp. Sci. pp. 98-108.

Chomsky, N. (1963) Formal properties of grammars. Handbook of Mathematical Psychology 2, John Wiley and Sons, pp. 323-418.

Church, A. (1941) The Calculi of Lambda-Conversion. Annals of Mathematics Studies 6, Princeton.

Cook, S.A. (1971) The complexity of theorem proving procedures. Proc. 3rd Ann. ACM Symposium on Theory of Computing, pp. 151-158.

Cook, S.A. and R.A. Reckhow (1974) On the lengths of proofs in the propositional calculus. Proc. 6th Ann. ACM Symposium on Theory of Computing, pp. 135-148.

Cook, S.A. (1980) Towards a Complexity Theory of Synchronous Parallel Computation. Presented at Internationales Symposium uber Logic and Algorithmik zu Ehren von Professor Ernst Specker, Zurich, Switzerland, February.

Cook, S.A. and R.A. Reckhow (1973) Time bounded random access machines. J. Comput. System Sci. 7:4, 354-375.

Davis, M. and J.W. Elaine (1983) Computability, Complexity and Languages. Academic Press.

Dymond, P. and S.A. Cook (1980) Hardware complexity and parallel computation. 21st Ann. IEEE Sympos. Found. Comput. Sci. pp. 360-372.

Dymond, P. (1984) Ph.D. thesis, Department of Computer Science, University of Toronto.

Fischer, M.J. and M.O. Rabin (1974) The complexity of theorem proving procedures. Project MAC Report, MIT, Cambridge, Mass.

Fortune, S. and J. Wyllie (1978) Parallelism in random access machines. Proc. 10th ACM Sympos. Theory of Comput. pp. 114-118.

Garey, M.R., D.C. Johnson and L. Stockmeyer (1974) Some simplified polynomial complete problems. Proc. 6th Ann. ACM Symposium on Theory of Computing, 47-63.

Garey, M.R. and D.S. Johnson (1979) Computers and Intractability: a Guide to the Theory of NP-Completeness. Freeman, San Francisco.

Gill, J.T. (1974) Computational complexity of probabilistic Turing machines. Proc. 6th Ann. Symposium on Theory of Comput.

Gödel, K. (1931) Über formal unentscheidbare Sätze der Principia Mathematica und verwandter Systeme I. Monatshefte fur Mathematik und Physik 38, 173-198.

Goldschlager, L. (1978) A unified approach to models of synchronous parallel machines. Proc. of 10th Ann. IEEE Sympos. Theory of Comput. San Diego, California, pp. 89-94.

Hartmanis, J., P.M. Lewis and R.E. Stearns (1965) Classifications of computations by time and memory requirements. IFIP International Congress 1, 31-35, Spartan Books.

Hartmanis, J. (1971) Computational complexity of random access stored program machines. Mathematical Systems Theory 5:3, 232-245.

Hong, J.W. (1980) On some deterministic space complexity problems. Proc. of 12th Ann. Symp. on Theory of Comp. pp. 310-317 and SIAM on Comp. 11 (1982).

Hong, J.W. (1980) On similarity and duality of computation. Proc. of 21st Ann. IEEE Sympos. Found. Comput. Sci. pp. 348-359.

Hong, J.W. (1984) The complexity of formal proving. Scientia Sinica, Ser. A, 27:10, 1046-1054.

Hong, J.W. (1984) On similarity and duality of computation (I), Information and Control. Vol. 62 nos. 2/3 (August/September), pp. 109-128.

Hong, J.W. (1984) A tradeoff theorem for space and reversal. Theoret. Comput. Sci. 32, 221-224.

Knuth, D.E. (1968) The Art of Computer Programming Vol. I: Fundamental Algorithms, Addison-Wesley, Reading, Mass.

Knuth, D.E. (1973) The Art of Computer Programming Vol. III: Sorting and Searching, Addison-Wesley, Reading, Mass.

Karp, R.M. (1972) Reducibility among combinatorial problems. In Miller and Thatcher (1972) Complexity of Computer Computions. Plenum Press, pp. 85-104.

Kleene, S.C. (1936) General recursive functions of natural numbers. Math Annalen 112, 340-353.

Kleene, S.C. (1952) Introduction to Metamathematics. Van Nostrand, Princeton.

Kleene, S.C. (1956) Representation of events in nerve nets and finite automata. In Automata Studies (Shannon and McCarthy, eds.). Princeton University Press, pp. 3-40.

Lewis, H.R. and C.H. Papadimitriou (1980) Symmetric space-bounded computation. TR-08-80, Aiken Computation Laboratory, Havard University.

McCarthy, J. et al. (1965) LISP 1.5 Programmers manual. MIT Press, Cambridge, Mass.

Meyer, A.R. and L. Stockmeyer (1973) Nonelementary word problems in automata and logic. Proc. AMS Symposium on Complexity of Computation, April 1973.

Post, E.L. (1943) Formal reductions of the general combinatorial decision problem. Am. J. of Math. 65, 197-268.

Papadimitriou, C.H. and K. Steiglitz (1982) Combinatorial Optimization: Algorithms and Complexity. Prentice-Hall, Englewood Cliffs, N.J.

Pippenger, N. (1979) On simultaneous resource bounds. Proc. 20th Ann. IEEE Sympos. Found. Comput. Sci., October, pp. 307-311.

Pratt, V. and L. Stockmeyer (1978) A characterization of verter machines. J. Comput. System Sci. 12, 198-211.

Rabin, M.O. and D. Scott (1959) Finite automata and their decision problems. IBM Journal of Research and Development 3:2, 114-125 (April 1959).

Rabin, M.O. (1963) Probabilistic automata. Information and Control 6, pp. 230-245 (September 1963).

Reif, J.R. (1980) On probabilistic and symmetric parallel computation. TR-22-80, Aiken Computation Laboratory, Havard University.

Rogers, H.J. (1967) Theory of recursive functions and effective computability. McGraw-Hill.

Ruzzo, M.L. (1979) On uniform circuit complexity. Proc. 20th Ann. IEEE Sympos. Found. Comput. Sci., October, pp. 312-318.

Savitch, W. and Stimmson, M. (1978) Time Bounded random access machines with parallel processing. J. Assoc. Comput. Mach. 26, 103-118.

Savitch, W.J. (1970) Relationship between nondeterministic and deterministic tape complexities. J. Comput. System Sci. 4:2, 177-192.

Schonhage, A. (1979) Storage Modification Machines. Technical report, Mathematisches Institute, Universitat Tubingen, Germany.

Sedgewick, R. (1983) Algorithms. Addison Wesley, Reading, Mass.

Stockmeyer, L.J. and A.R. Meyer (1973) Word problems requiring exponential time. Proc. 5th Ann. ACM Symp. on Theory of Computing.

Thompson, C.D. (1979) Area-time complexity for VLSI. 11th ACM Sympos. Theory of Computing, pp. 81-88.

Turing, A.M. (1936) On computable numbers, with an application to the Entscheidungsproblem. Proc. London Mathematical Soc. Ser. 2, 42, 230-265. Corrections, ibid., 43 (1937), 544-546.

Turing, A.M. (1950) Computing machinery and intelligence. Mind 59 (n.s. 236) 433-460.

Tarjan, R.E. (1972) Depth first search and linear graph algorithms. SIAM J. Computing 1:2, pp. 146-160.

Yap, C.K. (1987) Introduction to the theory of Complexity Classes. Oxford University Press (in press).

Index

accept   6, 19
address   81
   direct a.   81
   indirect a.   81
adjacency matrix   4
Adleman, L.M.   203
Aho, A.V.   76, 151
alphabet   2
   input a.   5
   output a.   5
alternating   76, 175

Boolean function   165
Boolean operation   103, 136, 142, 165
Borodin, A.   76, 139
boundary theorem   210
BPS   222
branch   23
Brent, R.P.   76, 156

Cartesian product   1
Chandra, A.K.   175, 217
choices, sequence of   159
Church, A.   18, 44
Church-Turing Thesis   18, 44
CLIQUE   208
closure   1, 3
   Kleene c.   3
   positive Kleene c.   3
COLOURABILITY   209

complete   205
   C-c.   205
complete problem   205
complexity   44, 76
   choice c.   169, 182
   c. class   194, 195
   computational c.   44
   reversal c. $r(n)$ 48, 169, 181
   space c. $s(n)$ 47, 169, 182
   time c. $t(n)$ 47, 169, 182
composition (of Turing machines)   22
computability   17, 44
computation   17, 44
   c. graph   177
   parallel c.   79
   sequential c.   79
   c. tree   167, 183
computational model   17, 76, 102, 136
   parallel c.m.   79, 102, 136
   real c.m.   153
   sequential c.m.   79, 102
computational type   76, 166, 185
   general c.t.   185
   logical c.t.   165
   non-trivial c.t.   170
concatenation   2, 9
concurrent reading (CR)   145
concurrent writing (CW)   145
configuration   59
   phase c.   60
   s-c.   162

constructible   155
Cook, S.A.   76, 80, 195, 207
cost
   logarithmic c.   86
   uniform c.   86
CRCW   146
CREW   146
cycle detecting problem   52

decidable   40
depth   136, 220
description   85
deterministic   5, 76
   non-d. state   168, 169
difference   1
duality   210
Dymond, P.   63, 76, 77

edge   3
equivalent
   e. logical types   170
   e. normal systems   218
exclusive reading (ER)   218
exclusive writing (EW)   146

fan-in   3
fan-out   3
finite automaton   5
   deterministic $f.a.$(DFA)   5
   non-deterministic $f.a.$(NDFA)   7
   non-deterministic $f.a.$ with
     Λ-moves   8
   two-way deterministic $f.a.$(2DFA)   14
finite state control (FC)   5, 18
Fisher, M.J.   77
flat case   75
formula   218

gate   136, 142
   input g.   136, 142
   input value g.   151
Gödel, K.   217
Gödel's incompleteness theorem   217
Goldschlager, L.M.   76
graph   3
   composed g.   70
   deterministic g.   125
   directed g.   3
   m-multilevelled g.   72
   reversed g.   4
   undirected g.   4

hard   205
   C-h.   205
hardware modification machines (HMM)
   76, 156
heap   92
homomorphism   12, 13
Hopcroft, J.E.   76, 151

input form   81, 82
   first $i.f.$ 81
   second $i.f.$ 82
   third $i.f.$ 82
instantaneous description (ID)   14, 19, 85
instruction   81, 104, 145
   admissible i.   81
intersection   1

join   71

Karp, R.   208
Kleene, S.C.   3, 11

Kung, H.T.   76, 156

language   2
leaf labeled binary tree   185
length   220
Lewis   204
linearly similar   169, 170
linearly simulate   169, 170
linear speed-up theorem   53, 54
log-space constructible   68, 137, 143
   l.s.c. codings   68
   l.s.c. functions   68
   l.s.c. graphs   69
   l.s.c. relations   70
log-space transform machine (LSTM)
   65, 66
loop   23
lower bound   49

machine
   (f,g)-$\underline{m}$.   78
matrix transpose   118
minimalization   32
Model (R,S)   78

narrow case   75
NC   67, 81, 195, 215
next move function   19, 159
nice function   76
nice pair of functions   73
nice triple of functions   152
normal system   218
NP   195
NPSPACE   195

oblivious   58

output function   5

Papadimitriou   204
partial function   32
path   4
phase   48, 66, 84
Pippenger, N.   58, 67, 77, 142
polynomially comparable   152
polynomially related   45
Post, E.   219
Post system   219
power set   1
PPS   222
PRAM   144
$PRAM_o$ 145
Pratt, V.   76, 102, 125, 132
prefix   2
   proper $\underline{p}$.   2
premise   218
primitive recursion   27
processor   144, 152
production   221
program   81, 104, 144
proof   220
propositional calculus   218
PSPACE   195
pumping lemma   11

Rackoff, C.   63
RAM_o 86
random access machine (RAM)   76, 80
   multi-index $\underline{r.a.m.}$ 80, 81
   parallel $\underline{r.a.m.}$ (PRAM)   76, 144
reasonable coding   39
Reckhow, R.A.   76, 80
recursive language   40